Quantum Physics

KNOWLEDGE
IN A
NUTSHELL

Quantum Physics

KNOWLEDGE IN A NUTSHELL

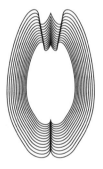

Sten Odenwald

SIRIUS

SIRIUS

This edition published in 2020 by Sirius Publishing, a division of
Arcturus Publishing Limited,
26/27 Bickels Yard, 151–153 Bermondsey Street,
London SE1 3HA

Copyright © Arcturus Holdings Limited

ISBN: 978-1-78950-583-2
AD006696UK

Printed in Singapore

Contents

Introduction

Quantum physics is perhaps one of the most challenging and mind-numbing subjects in all of physics. It is widely understood by physicists themselves as one of the deepest subjects in the scientific study of our physical world. Even Richard Feynman, noted that 'I think it is safe to say that no one understands quantum mechanics. Do not keep saying to yourself, if you can possibly avoid it, "but how can it be like that?" because you will go down the drain into a blind alley from which nobody has yet escaped. Nobody knows how it can be like that'.

This book will try to describe the highlights of this deep understanding of matter and forces starting from the Atomist School of ancient Greece through the most advanced, current experiments on the nature of matter. The plethora of books that present quantum mechanics in a popular writing style bespeak a reoccurring fascination with this topic, and there are many different styles to choose from. This has been a great benefit to the novice who is trying to obtain a snapshot of the key issues and discoveries in this field. Because quantum physics is an evolving subject that benefits every year from new experimental studies, it is most certainly important to update the public discussion of this field. Even books written as little as five years ago are out-of-date in certain key areas of investigation, most notably in the areas of string theory, quantum gravity and issues of quantum measurement.

Unlike most other areas in science, physics can only be expressed accurately in mathematical terms. I will avoid the extravagant use of mathematical equations, and the very few that are presented should be regarded as paintings in an art museum to be admired for their form much as one might admire a painting without delving deeply into its subject matter or the artist's paintbrush strokes. The compromise, however, is that numerical information may in some instances be calculated to show 'how it is done' and to set the scale for the phenomena being described.

Quantum physics also uses a plethora of terminology that are often intimidating to the non-physicist. Unfortunately it is impossible to write about quantum physics without mentioning 'wave functions', 'fermions' and the many terms such as 'SU(3)' that have appeared in the subject areas of Grand Unification Theory and quantum measurement. There is no good solution to how to discuss important concepts without mentioning the specific objects being described, so the reader is encouraged to consider these terms when they appear as proper names such as 'Steven' or 'Mary' that themselves require no further definition.

The entire rubric of quantum physics, which will be described in this book in some detail, can be distilled into a small list of conclusions about the physical world, which I attempt to summarize at the end of each chapter. Each of these conclusions are in themselves worthy of deep contemplation and lead to a dramatic transformation in the way we must view our physical world. The atomic matter from which our universe is built does not have a definite property to it until the instant of observation or measurement, and changes instantaneously between a particle and a wave-like description depending on the experimental question being asked. What we consider to be 'empty space' as a vacuum is an illusion, as are the very natures of time and space. Finally, the ingredients to the universe behave in some sense as though they require the act of observation to bring them into existence. Finally, the indeterminacy of the quantum description to the atomic world is not an artefact of missing information that could be provided by some new, larger theory of the world. After only 100 years of investigation and experimentation, we are only a bit further ahead in understanding why the world is built the way it is. Nevertheless, what we have learned by 2019 most assuredly sets the stage for entirely new investigations in the years to come.

CHAPTER 1

The Dawn of Atomic Physics

The idea that the things around us, such as rocks, air and water, are themselves made from other ingredients more elementary than the physical world visible to us is one that was long in coming. From the dawn of the written word – fashioned from Sumerian cuneiform, Egyptian hieroglyphics, or even ancient Chinese pictograms – there is scarcely a hint that our very distant ancestors considered such ideas noteworthy. Of course, humans at all times could distinguish among the many diverse substances: from rocky compounds and elementary gold to the bewildering variety of biological matter in trees, plants and animals. But the idea that there were still more elementary substances behind the ones we commonly see was inconceivable. So far as recorded history can tell us, this all changed during the last few centuries BCE, when ancient Indian philosophers and ancient Greek proto-scientists came up with the idea of the 'smallest particles of matter'. The Indian idea is credited to Acharya Kanad, who lived sometime between the 6th and 2nd centuries BCE in Prabhas Kshetra (near Dwaraka) in Gujarat, India. Beginning with grains of rice, he developed the idea of the smallest particles of matter, which he later called *Parmanu*. Kanad later founded the Vaisheshika school of philosophy, where he taught his ideas about the atom and the nature of the universe. Virtually identical ideas about the nature of matter were developed, either independently or by the natural diffusion of ideas, by the ancient Greeks.

THE ATOMIC UNIVERSE

The theory of Atoms (from the Greek ἄτομος, atomos, meaning indivisible), proposed by Democritus and Leucippus in the 5th century BCE was a dramatic departure from prevailing ideas around the Mediterranean and in the Far East, in which matter was a continuum characterized by a handful of distinct properties. The Atomists made the innovative proposal that there were an infinite number of atoms varying in shape, and that these shapes determined the properties of the matter they combined to create. For example, iron atoms had tiny hooks that linked together at

Democritus and Leucippus developed the ancient Greek idea of the atom.

room temperature, making it a strong solid. Atoms were so small they could not be seen, and like marbles in a half-empty box, they could rearrange themselves to form new substances.

This meant that in addition to atoms there must also exist a completely empty Void, otherwise these atoms would be locked together into a vast immovable mass for eternity. Only the Void made atoms a workable construct, and so from this time forward, the properties of empty space and atoms were intimately linked together. In 55 BCE, Lucretius, in his *On the Nature of Things*, described the shapes of atoms as follows:

'*Since atoms are what they are by nature, and not cut by hand to a single predetermined pattern, some of them must have shapes unlike some others. By reasoning, we may readily comprehend why lightning-fire is much more penetrating than ours that comes from torches here on earth. Or one may say that the lightning fire is finer and made in smaller shapes, and thus can pass through opening that our fire cannot, spring as it is of wood and torch.... Things which are hard and dense must be composed of particles hooked*

and barbed and branch-like, intertwined
and tightly gripped.'
(*Lucretius, Epicurean and Poet*, by John Masson 1907)

Meanwhile, several hundred years after the Atomists established their fledgling school of thought in the West, Aristotle (384–322 BCE) espoused the existence of only five essences: Earth, Air, Fire, Water and Aether. In fact, he found the idea of the Atomist's empty Void abhorrent, and as one of the most influential philosophers of his time he was able to mount many criticisms of the Atomists' school. Relatively quickly the theory of Atoms went into eclipse as a theory of matter. Nevertheless, by the 2nd century CE, alchemists, metallurgists and jewellers knew of a great number of elemental species of Aristotle's Earth element and recognized that some of these, such as gold, were unable to mix with other compounds, making them more elementary than other forms of Earth. There were strong dividing lines between what the Aristotelian theoreticians were saying, and what was known by alchemists experimentally. By the 14th century, there was a revival of interest in the Atomist school of thought, but because atomism was associated with the philosophy of Epicureanism, which contradicted orthodox Christian teachings, belief in atoms was not considered acceptable.

Another long period of time elapsed before chemist Robert Boyle (1627–1691) and Sir Isaac Newton (1642–1727) defended the concept of atomism. Once again this idea of the structure of matter reappeared within the growing scientific community, nearly 2,000 years after it had first been proposed. In the *Philosophiae Naturae Principia Mathematica*, Newton claimed '...the least parts of bodies to be—all extended, and hard and impenetrable, and moveable, and endowed with their proper inertia'. Newton also went on to propose that atoms interacted among themselves through non-gravitational forces. However, his atoms were rather featureless and did not account for any of the chemical properties of atoms, only in their densities being different to account for

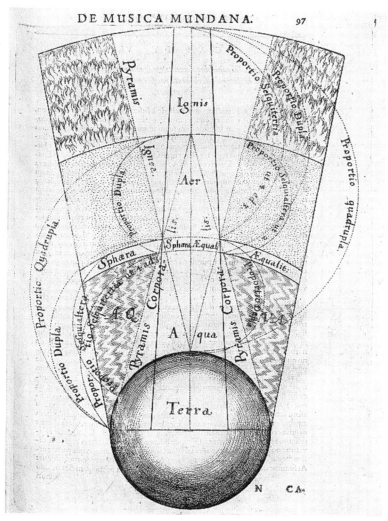

Aristotle's four elements, earth, air, fire and water held sway as the basic elements of matter until the 14th century.

water being lighter than rock. So, while tremendous advances were being made in studying the movement of bodies under the influence of gravity, our basic understanding of the nature of matter remained extremely primitive. The fleshing out of the

Antoine Lavoisier discovered the chemistry of oxygen.

atomic model proceeded very rapidly once chemists became more analytical about how atoms and molecules combined and transformed in a variety of reactions they could observe at the laboratory bench.

ANALYTICAL CHEMISTRY AND THE PROPERTIES OF ATOMS

Alchemy and what we now call chemistry were a cottage industry in both the Eastern and Western worlds as people struggled to synthesize gold from baser compounds. To find just the right mixture and process to make gold appear at the bottom of their crucibles required enormous amounts of time, and careful notations about what was tried and what failed. Along the way, certain compounds and substances began to appear rather regularly out of these alchemical syntheses. Some alchemists ignored these results and continued with their fevered search for various life-extending elixirs and gold, while others became increasingly intrigued by how some specific combinations led to specific end results. In 1777, Carl Scheele (1742–1786) was the first to recognize that the ordinary air described by Aristotle was itself a compound substance consisting of foul air (nitrogen) and fire air (oxygen). No doubt, foul air had trace impurities of nitrogen sulphide in it, which gave it the rotten eggs aroma. The English chemist Joseph Priestly (1733–1804), meanwhile, had succeeded in producing a variety of gases under laboratory conditions. One of these was called dephlogisticated air, which we now know as oxygen.

In France, Antoine Laurent Lavoisier (1743–1794) announced in 1783 that combustion was a chemical process involving the combination of Priestly's newly discovered oxygen with the combustible substance. Following Lavoisier's investigations, Earth, Air, Fire, Water and Aether were no longer regarded as the fundamental building blocks of nature, and it was understood that they could, themselves, be composed of yet more fundamental substances. Lavoisier's *Traité élémentaire de chimie* (*Elementary Treatise of Chemistry*), published in 1789 a few years before his beheading at the height of the French Revolution, is regarded as the first modern textbook of chemistry and included the first systematic listing of some 33 elementary substances known by his time.

By the end of the 18th century we had therefore entered a period where atoms were being considered for their pragmatic efficiency in helping to classify elementary substances, though

	Noms nouveaux.	Noms anciens correſpondans.
	Lumière.	Lumière.
	Calorique.	Chaleur. / Principe de la chaleur. / Fluide igné. / Feu.
Subſtances ſimples qui appartiennent aux trois règnes & qu'on peut regarder comme les élémens des corps.	Oxygène.	Matière du feu & de la chaleur. / Air déphlogiſtiqué. / Air empiréal. / Air vital. / Baſe de l'air vital.
	Azote.	Gaz phlogiſtiqué. / Mofète. / Baſe de la mofete.
	Hydrogène.	Gaz inflammable. / Baſe du gaz inflammable.
Subſtances ſimples non métalliques oxidables & acidifiables.	Soufre.	Soufre.
	Phoſphore.	Phoſphore.
	Carbone.	Charbon pur.
	Radical muriatique.	Inconnu.
	Radical fluorique.	Inconnu.
	Radical boracique.	Inconnu.
Subſtances ſimples métalliques oxidables & acidifiables.	Antimoine.	Antimoine.
	Argent.	Argent.
	Arſenic.	Arſenic.
	Biſmuth.	Biſmuth.
	Cobolt.	Cobolt.
	Cuivre.	Cuivre.
	Etain.	Etain.
	Fer.	Fer.
	Manganèſe.	Manganèſe.
	Mercure.	Mercure.
	Molybdène.	Molybdène.
	Nickel.	Nickel.
	Or.	Or.
	Platine.	Platine.
	Plomb.	Plomb.
	Tungſtène.	Tungſtène.
	Zinc.	Zinc.
Subſtances ſimples ſalifiables terreuſes.	Chaux.	Terre calcaire, chaux.
	Magnéſie.	Magnéſie, baſe du ſel d'Epſom.
	Baryte.	Barote, terre peſante.
	Alumine.	Argile, terre de l'alun, baſe de l'alun.
	Silice.	Terre ſiliceuſe, terre vitrifiable.

Lavoisier's list of elements c. 1789.

there was still hardly much detail as to what they were in and of themselves. The English chemist John Dalton (1766–1884) is usually credited with proposing the atomic theory for the elements and their chemical reactions. His basic ideas were:

1. Elements are made of extremely small particles called atoms.
2. Atoms of a given element are identical in size, mass and other properties; atoms of different elements differ in size, mass and other properties.
3. Atoms cannot be subdivided, created or destroyed.
4. Atoms of different elements combine in simple, whole-number ratios to form chemical compounds.
5. In chemical reactions, atoms are combined, separated or rearranged.

Dalton may have been influenced by the Irish chemist Bryan Higgins (1741–1818), who was one of the first chemists to inquire as to the internal structure of atoms. He came up with the brilliant idea that an atom consisted of a heavy central particle surrounded by an atmosphere of a material called caloric, which was the supposed

John Dalton developed the first internal model of elementary atoms.

substance of heat at the time – the size of the atom being determined by the diameter of the caloric atmosphere. Dalton's subsequent atomic theory was similar, but was focused more upon his detailed measurements of chemical interactions from which he deduced a Law of Multiple Proportions: If two elements form more than one compound between them, then the ratios of the masses of the second element which combine with a fixed mass of the first element will be ratios of small whole numbers. This led, in turn, to the idea that the law could be most easily explained as the interaction of atoms of definite and characteristic weight. At this level, chemistry was not about the modern ideas of electrons and their interactions, but was simply the relationships between the differing masses of the elementary atoms. Higgins' idea, however, would eventually lead to the idea that it wasn't the heavy nucleus which was involved in chemistry, but some other circum-nuclear aspect to the atom – revealed, nearly a century later, as the electron cloud. We know about Dalton's atomic theory because it was published widely.

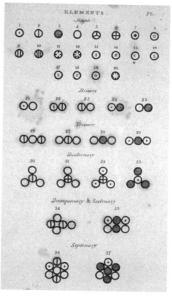

Atoms and molecules shown in Dalton's A New System of Chemical Philosophy (1808).

Higgins' theory never received widespread publication, primarily because Higgins' suffered from declining a mental condition in 1803.

During the late 1860s, Dmitri Mendeleev (1834–1907) and Julius Meyer (1830–1895) developed a classification scheme for the known elements. Following Lavoisier and Dalton's tabulations, over 37 new elements had been added since the turn of the 1800s, with the total known elements increased to over 58 elements by Mendeleev's time. The Mendeleev-Meyer table was divided into 9 columns called groups and 16 rows called periods, such that the groups represented elements with similar chemical properties and the rows were successively heavier elements of the same chemical properties. Mendeleev used this scheme to predict the existence of three new elements that he called eka-boron, eka-aluminium and eka-silicon, which were later discovered and renamed scandium, gallium and germanium.

Tabelle II.

Reihen	Gruppe I. — R²O	Gruppe II. — RO	Gruppe III. — R²O³	Gruppe IV. RH⁴ RO²	Gruppe V. RH³ R²O⁵	Gruppe VI. RH² RO³	Gruppe VII. RH R²O⁷	Gruppe VIII. — RO⁴
1	H=1							
2	Li=7	Be=9,4	B=11	C=12	N=14	O=16	F=19	
3	Na=23	Mg=24	Al=27,3	Si=28	P=31	S=32	Cl=35,5	
4	K=39	Ca=40	—=44	Ti=48	V=51	Cr=52	Mn=55	Fe=56, Co=59, Ni=59, Cu=63.
5	(Cu=63)	Zn=65	—=68	—=72	As=75	Se=78	Br=80	
6	Rb=85	Sr=87	?Yt=88	Zr=90	Nb=94	Mo=96	—=100	Ru=104, Rh=104, Pd=106, Ag=108.
7	(Ag=108)	Cd=112	In=113	Sn=118	Sb=122	Te=125	J=127	
8	Cs=133	Ba=137	?Di=138	?Ce=140	—			— — — —
9	(—)		—	—	—		—	
10	—	—	?Er=178	?La=180	Ta=182	W=184	—	Os=195, Ir=197, Pt=198, Au=199.
11	(Au=199)	Hg=200	Tl=204	Pb=207	Bi=208	—		—
12	—	—	—	Th=231	—	U=240	—	— — — —

der chemischen Elemente.

Mendeleev's 1871 periodic table. The eight groups represent elements with the same chemical reactions, while the 12 Periods represent elements with decreasing atomic radius and increasing ionization energy along each row.

This periodic table of the elements was modified in 1894 by Lord Rayleigh (1842–1919) and Sir William Ramsay (1852–1916) who included a separate column for the inert gases: helium, neon, argon, krypton and xenon. Nevertheless, even by the end of the 19th century, the groupings of the elements and their systematic properties were still seen as a feature of familiar, though perplexing, properties of their intrinsic density and mass. In some corners of the physics community, the very idea of atoms of matter was still not settled. Ernst Mach (1838–1916), for example, believed that atoms were nothing more than a theoretical construct. On the other hand, Ludwig Boltzmann (1844–1906), one of the founders of the atom-based kinetic theory of gas and thermodynamics, considered them to be absolutely real.

THE ELECTRON

While chemists had scored enormous successes in classifying the elements in a systematic fashion, there were no hints as to the structures of these various elementary atoms that allowed them to be so mathematically organized by atomic mass and chemical affinities. Along a parallel but largely independent track came the discoveries of new 'subatomic' particles, the first of these being the electron.

Static electricity had been known since the time of the ancient Greeks, but it wasn't until the 18th century that investigators such as the French physicist Charles François du Fay (1698–1739) and the American polymath Benjamin Franklin (1706–1790) determined that a single electrical fluid of charged particles was responsible for most electrostatic events including lightning. By 1874, Irish physicist George Johnstone Stoney (1826–1911) suggested that there existed a single definite quantity of electricity. He was able to estimate its value, and also gave name to this so-called electric ion, writing in a 1894 article 'Of the "Electron" or Atom of Electricity' in the *Philosophical Magazine*: '…an estimate was made of the actual amount of this most remarkable fundamental unit of electricity, for which I have since ventured to suggest the name electron.' But the expectation was that this could never be found outside the atom and was permanently attached to it.

The discovery of cathode rays by J. W. Hittorf in 1869 (1824–1914) showed that atoms could also be made to emit negatively charged particles that behaved like electricity. The internal structure of atoms now had to include electrons. These experiments and others quickly led to a model for the atom, which ultimately provided a framework for understanding the regularity

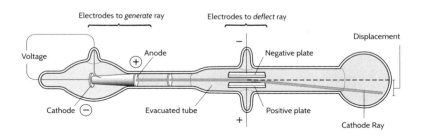

A simple cathode-ray tube used by Hittorf to discover free electrons.

of the Periodic Table as well as some of the new advancements in the growing science of spectroscopy.

THE PLUM PUDDING MODEL

The model was first proposed in 1904 by Sir Joseph John Thomson (1856–1940), who thought that atoms contained a uniform mixture of positive and negative 'corpuscles'. George FitzGerald (1851–1901) had suggested in 1900 that the atom owed its entire mass to a large population of electrons. For instance, the hydrogen atom contained 500 electrons and oxygen contained 8,000.

The 'Plum Pudding' model of the atom proposed by J. J. Thomson.

THE RUTHERFORD ATOM

Experiments by Henri Becquerel (1852–1908) with a compound called potassium uranyl sulphate ($K_2O_{10}S_2U$) showed that the element uranium in this compound was emitting particles. Ernest Rutherford (1871–1937) continued these investigations and discovered that some active elements emitted what he classified as alpha, beta and gamma particles. The alpha particles were later identified as nuclei of the helium atom, while beta particles were electrons, and gamma particles were a form of high-energy light.

Rutherford came up with an ingenious idea for probing the structure of atoms. He would use the alpha particles as bullets and let them pass through a thin screen of gold foil. If atoms were a uniform ball of mixed charges, the alpha particles would pass through the screen unaffected. But what he saw was that some alpha particles were reflected or scattered through the foil at very high angles to their initial path. Rutherford's experiments revealed that, instead of Thomson's uniform mixture of positive and negative corpuscles, the atom had a heavy, compact nucleus surrounded by a cloud of electrons, and in 1911 his model of the atom finally appeared in print. The positive nuclear charge was

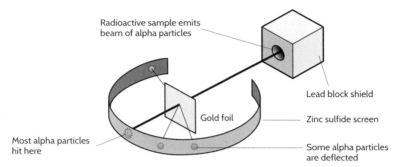

Radioactive sample emits
beam of alpha particles

Lead block shield

Gold foil

Zinc sulfide screen

Most alpha particles
hit here

Some alpha particles
are deflected

Rutherford's scattering experiment reveals the nucleus of the atom.

eventually identified as the proton by Rutherford, and by 1931, a heavy neutral companion particle was discovered by Irène Joliot-Curie (1897–1956) and James Chadwick (1891–1974) and called the neutron.

This process of atoms emitting other forms of matter called radioactivity signalled the fall of the last recognizable element of the ancient Atomist school. After more than 2,000 years, scientists had surpassed the thinking of Greek and Indian philosophers and found their way to a completely new understanding of the composition of matter: atoms were indeed real features of the physical world, but they were not eternal. This turned out to be only the beginning of a larger story that was to unfold during the first decades

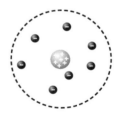

The initial Rutherford atom showing a dense nuclear core surrounded by electrons.

of the 20th century. The search was now on to understand these subatomic particles, and the first ones to garner the most interest were the electrons themselves.

ATOMIC STRUCTURE REVEALED BY SPECTROSCOPY

The discovery that atoms possessed internal constituents came as a result of the development of a new technique for analyzing light pioneered by Newton himself: spectroscopy. Early in the 19th

Fraunhofer demonstrating his new diffraction-grating spectroscope.

century, William Wollaston (1766–1828) and, in 1814, Joseph von Fraunhofer (1787–1826) had independently improved on Newton's prismatic technique for forming a spectrum of sunlight, and discovered dark bands against the rainbow colours of the sun's spectrum – Wollaston in 1802 and Fraunhofer in 1814, with his invention of the spectroscope. Although Wollaston did not consider the handful of spectral lines he spotted to be scientifically significant, Fraunhofer recognized at once their unique characteristics and ultimately catalogued over 574 individual dark lines.

Fraunhofer was a master optician, and his instruments were in high demand because of their optical performance. However, he realized early on that to base optical performance on solar 'white light' alone would lead to inaccuracies in optical fabrication. So, he explored creating light in only one colour via prismatically dispersing the optical spectrum and using the monochromatic, filtered light to test his optical designs. Instead of using a pinhole in a screen as a source of sunlight, which Newton had done, he chose to use a slit, which was the missing ingredient needed to image the individual dark lines. He quickly evolved this method by eliminating the prism and using a diffraction grating to disperse the light in wavelength. Despite having discovered these lines, Fraunhofer did not really understand or worry about their origins. This step required another 30 years beyond his death to gain any further scientific interest.

JOSEPH VON FRAUNHOFER

Orphaned at the age of 11, Joseph von Fraunhofer (1787–1826) worked in a glass house as a young apprentice until it burned down. He attracted two wealthy benefactors who encouraged him to continue his studies, and became adept at making high-quality optical instruments. By the age of 32 he had become the Director of the Optical Institute in Benediktbeuern, and he single-handedly made Bavaria the centre of high-precision optical instrument design and manufacture. He died at age 39 from the chronic inhalation of heavy metal vapours, a common cause of death for glassmakers of the time.

By 1859, Gustav Kirchhoff (1824–1887) and Robert Bunsen (1811–1899) were able to show that each element had its own unique combination of spectral lines. As specific elements were burned in a candle flame, they produced a pattern of lines that matched some of the solar Fraunhofer lines, specifically the D-line of sodium. Kirchhoff experimented with sources of light such as electrical arcs and found that the bright lines in the sodium spectrum not only appeared at the same wavelength as some of the dark lines produced by the sun, but the dark lines actually absorbed the light from the corresponding bright lines. This led to Kirchhoff's Law, which stated that the emitted and absorbed power of lines from all elements at the same wavelength were equal. This immediately opened up the investigation of the solar spectrum in terms of matching the Fraunhofer absorption lines with emission lines created with elements available on Earth.

In 1862, these spectral analysis techniques were extensively applied to the analysis of the light from planets, stars and nebulae by William Huggins (1824–1910) and his neighbour, the chemist William Allen Miller (1817–1870). The French philosopher Auguste Comte had opined in 1835 that no one would ever surmise the chemical composition of the distant stars, but Huggins and Miller quickly proved this expectation to be false. What they found was that the planets possessed gases similar to earth but with some significant differences. The star Aldebaran, for example, had a variety of lines corresponding to the elements hydrogen, sodium, magnesium and calcium, but Betelgeuse and Beta Pegasi contained no hydrogen at all.

An intriguing feature of the spectra from some elements is that the spectral lines often occur in sequences. By 1871 George Stoney (1826–1911) went on to propose that one periodic motion in the molecule of the incandescent gas may be the source of a whole series of lines in the spectra of the gas. This idea – that some numerical law underlies the regularity of the lines in the spectrum of hydrogen – was accepted for many years, but no one could succeed in producing a formula or explaining how it worked. Eventually in 1884, the schoolteacher and amateur numerologist Johann Balmer (1825–1898) through considerable trial and error deduced the formula:

$$\lambda = R \frac{n^2}{n^2 - m^2} \text{ nanometres}$$

Where R was a constant with an experimental value of 364.5 and m and n are the integers and each number pair (m,n) was associated with each specific spectral line at a wavelength of λ in the hydrogen atom. For example, with m=2 and n=3 one obtains $\lambda = 656.1$ nm which is close to the wavelength of the so-called Balmer-Alpha (B_α) line at 656.5 nm. As we will see in Chapter 3, this relationship was later refined by Johannes Rydberg (1854–1919) who determined what the constant R would be in terms of other physical constants based upon a simple physical model of the atom. He then applied the new Rydberg–Balmer formula to other single-electron atoms.

The discovery of numerical, and in particular integer-based, relationships among the spectral lines implied that, in addition to their chemical properties, atoms had another property that allowed them to emit only certain wavelengths of light. The integer nature of the m and n values ensured that each atom's lines would be found at only specific wavelengths and no others. The revolutionary nature of this mathematical realization is very hard to understate. Since the time of Galileo and Newton, physical systems could be described in terms of quantities such as mass (m), velocity (v), acceleration (a), which were continuous in nature. For instance, mass could come in any amount and be represented by continuous variable such as m for which any numerical value could be substituted. This led to smoothly-varying behaviour in the associated system with no discontinuous jumps. A ball rolling down an inclined ramp, or a planet orbiting the sun, did not move in a jerky manner but moved smoothly, and gained or lost energy in a similarly smooth manner. A Premier League football player could impart a kinetic energy of 100 Joules to a 0.4 kg football, but could also impart 101 Joules or even 100.00357 Joules because all numerical values are possible within the realm of what mathematicians call the Real Numbers. The discovery by Balmer and Rydberg that atomic spectral lines followed a formula involving integers and not the full gamut of the Real Numbers could only mean that, whatever was going on inside atoms required a non-Newtonian explanation for which none was as yet available. There were not an unlimited number of ways for an atom to emit light, unlike a Newtonian system for which all wavelengths should be possible for any one atom. The Balmer-Rydberg relationships

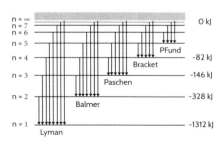

Atomic lines from the Balmer and Rydberg models. Lyman Series for n = 1, Balmer Series for n = 2, etc.

had to be answered by further research on just what was going on inside Dalton's atoms to make such emission possible.

One of the earliest proposals for physically explaining the Balmer–Rydberg relationship was by Lord Kelvin (1824–1907) in 1867 who devised the vortex model for the atom, in which the interior of the atom contained vibrating rings. These rings were loops of knotted ether, and the number and geometry of these atomic knots in the ether were the cause of the emission lines and their regularities.

Meanwhile, however, the prevailing theory of light was developed by James Clerk Maxwell (1831–1879) in his 1865 treatise 'A Dynamical Theory of the Electromagnetic Field'. This dealt with charged particles emitting continuous waves of 'electromagnetic' energy observed as light, and later detected by Heinrich Hertz (1857–1894) as radio waves – and there was nothing within Maxwell's Electrodynamics that gave any clue to a discrete emission mechanism. The combination of spectroscopy with discrete spectral lines, and Maxwell's continuous electromagnetism offered a contradictory perspective on atomic light emission that could not be resolved from within either theory alone.

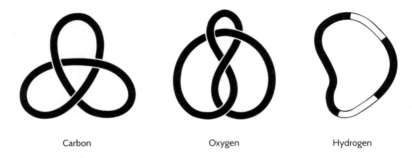

Carbon   Oxygen   Hydrogen

Lord Kelvin's knotted atoms attempted to explain the regularity of atomic lines due to integer-like features of atomic structure.

Key Points

- Early models of matter favoured basic continuous media, such as air, earth, fire and water, proposed by Aristotle.

- Atomism was created in India and in ancient Greece and involved minute, indestructible particles that carried the properties of their substance.

- The advent of spectroscopy witnessed a renewed interest in the internal structure of atoms to explain the periodic and narrow spectral lines, which served as atomic identifiers.

- Chemists discovered a vast number of elements beginning in the early 1800s, which were systematized into a periodic table of the elements by Mendeleev and Meyer.

- Several models for the atom arose from studies by chemists and physicists, including the homogeneous Plum Pudding Model with a mixture of positive and negative charges, and the Rutherford Model with a dense central nucleus surrounded by electrons.

- The production of narrow, single-frequency spectral lines by atoms could not be understood in terms of the known theories of the production of light, which was a continuous process, suggesting that atomic light was produced by some process that favoured new integer-based physical processes currently unknown by the end of the 19th century.

CHAPTER 2
Quantization

In addition to atomic structure, whose early development was described in Chapter 1, another area of research was destined to cause a profound change in the understanding of matter and light. The stimulus for this change was a deceptively simple phenomenon known to any blacksmith. As the temperature of a bar of iron increases, it continuously changes colour from red to orange, yellow and then white. It does this every time, and in exactly the same sequence of colours as the temperature increases. If such a regularity exists, surely some law of nature must circumscribe it?

THE BLACK BODY CURVE

The precise measurement of light intensity at different wavelengths was a significant technological challenge for much of the 19th century. In 1864, John Tyndall (1820–1893) used what he called a thermo-electric pile and attached this to a galvanometer to measure the intensity of light across a dispersed spectrum. He compared the colour of a heated platinum filament to the intensity of the light, which was later used in 1879 by Josef Stefan (1835–1893) to propose a law for heated surfaces in which their emission is proportional to the fourth power of their absolute temperature. From theoretical considerations, Ludwig Boltzmann derived a similar law in 1884, now known as the Stefan–Boltzmann Law. This law, which gives the flow of light energy through a surface (called flux) is customarily written as $F = \sigma T^4$, where F is the flux in watts/meter^2 and T is the black body surface temperature in kelvins. The Stefan–Boltzmann constant, σ,

John Tyndall was the first to quantitatively measure the intensity of light in the black body spectrum.

has the modern value of 5.670367×10^{-8} Watts meter$^{-2}\bullet$Kelvin^{-4}. This simple equation resulted in a major discovery in astronomy related to the temperature and luminosity of stars.

A star is a sphere of high-temperature plasma, which can be represented as having a radius, R, and a surface temperature, T. From the Stefan–Boltzman Law, the total light energy emitted by the star is simply $L = 4\pi R^2 F$, which with a little algebra becomes $L = 4\pi R^2 \sigma T^4$. This means that if you measure the solar temperature T using spectroscopic techniques, and its radius R using standard astrometric techniques, you can calculate its luminosity L in watts. For example, $T = 5770$ k and $R = 6.95 \times 10^8$ m results in $L = 3.8 \times 10^{26}$ watts. But because luminosity depends on the star's radius, you can now explain why two stars spectroscopically similar in temperature to the sun can have vastly different luminosities. For instance, Alpha Aquari has a temperature of 5,200 K, but a luminosity of 3,000 times our own sun because its radius is 77 times greater than the solar radius.

The actual, detailed shape of the spectrum from a heated body had to await the technology to make precise intensity measurements across a range of wavelengths. This level of precision became available in the 1890s and was used by Wilhelm Wien (1864–1928) to measure the radiation spectrum from a 'black body'. Prior to 1860, it was already known that the amount of radiation that a body emits was equal to the amount that it absorbs. For objects that were perfect absorbers of radiation, they would be exceptionally black in appearance, hence the name black body. In 1859, Gustav Kirchhoff (1824–1887) went one step further and proposed a universal spectral radiance function that was identical for every perfectly black body at every wavelength, which became known as Kirchhoff's Law of Radiation. In 1895, Wien created a black body by building an oven at a carefully regulated temperature, and then opening a very small hole into its interior through which the internal radiation could escape. The radiation was sent through a diffraction grating to spread out its wavelengths, and then a sensitive detector was scanned along the dispersed spectrum to register the intensity of the black body emission. From this and subsequent investigations, he was able to discern the shape of Kirchhoff's universal spectral radiance function.

The relationship between the temperature of the black body and the wavelength of the brightest light intensity was discovered by Wein in 1893 and this Wein

Gustav Kirchhoff developed the principle that a perfect absorber was also a perfect emitter at every wavelength.

Displacement Law was expressed as $\lambda = b/T$ where λ is the peak wavelength in millionths of a metre (a unit called the micrometre represented by the symbol μm), T is the absolute temperature in kelvins and the constant b has a value of 2898 μm-kelvins. For

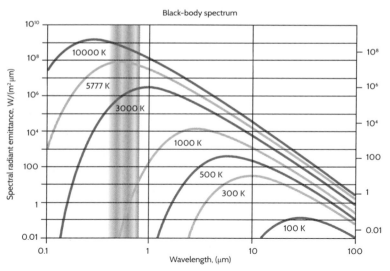

The intensity of a black body changes with its temperature which also causes its peak emission to shift to shorter wavelengths.

example, the light from the sun has a peak emission near 0.5 μm so its temperature is T = 2898/0.5 = 5796 kelvins.

A number of attempts were made to discern the mathematical form of this spectral radiance formula. Wein, himself, used thermodynamic arguments to propose a formula that looked like this in modern notation:

$$I\,(\lambda, T) = \frac{2hc^2}{\lambda^5}\ e^{-\frac{hc}{\lambda kT}}$$

where c is the speed of light, k is Boltzmann's constant and h is what later became Planck's Constant. Wein's formula worked reasonably well, but it failed to accurately predict the intensity of long-wavelength emission. Later between 1900 and 1905, Lord Rayleigh (1842–1919) and Sir James Jeans (1877–1946) proposed a new irradiance formula, the Rayleigh–Jeans Law, which was an approximation that seemed to work at long wavelengths. However, it suffered from significant departures at short wavelengths, termed the ultraviolet catastrophe.

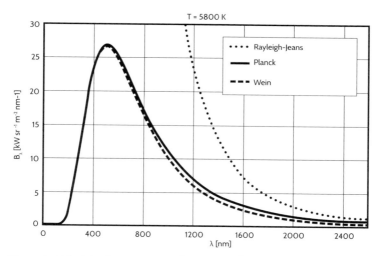

Computed black body spectrum vs. frequency ν for T = 5800K according to Planck's law. The Wein and Rayleigh-Jeans approximations are shown additionally as dashed lines.

$$I\,(\lambda,T) = \frac{2ckT}{\lambda^4}$$

So we are left with the equation provided by Wein that worked to calculate accurately the intensity of light from a black body for short wavelengths, but it failed for long wavelengths. Meanwhile, the Rayleigh–Jeans Law worked for long wavelength emission but failed for short wavelength light emitted by a black body. During the last decades of the 19th century, many physicists worked on this problem of finding a single mathematical equation for the black body curve, but it was not until 1900 that Max Planck (1858–1947) finally discovered the correct mathematical theory to explain this phenomenon. There was, however, a catch to this new-found knowledge. Because of the work by Maxwell on electrodynamics, previous theoretical arguments had assumed that light was a continuous phenomenon. When you mathematically counted the number of hypothetical emitters within a black body (called oscillators) to compute the emission at each wavelength, the oscillators were of unlimited number. This required the use of calculus, and specifically integrals over continuous variables, which produced a diverging sum as you made the calculation for shorter and shorter wavelengths. Planck decided to do away with this approach, but to get the mathematics to work correctly and produce a finite sum at short wavelengths, Planck invented the technique of 'quantization', treating light as though it came in packets, or quanta, of energy.

Each light quantum had an energy which was given by E = hn, so when Planck's Constant, h, whose value is 6.626 × 10^{-34} Joules-sec, was multiplied by the frequency of the light wave, the result was the amount of energy carried by the light quantum. For example, ordinary light with a frequency of 10^{15} cycles per second (called a Hertz: Hz) carries an energy of 6.6 × 10^{-19} Joules, while the light quanta responsible for radio waves at a frequency of 100 megahertz (MHz) carry 10 million times

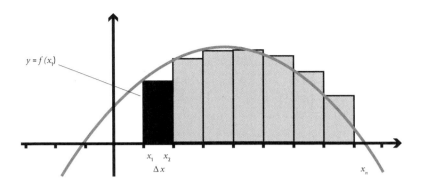

Planck's solution can be thought of as replacing a smooth sum provided by integral calculus with a discrete sum. The former assumption for the number of oscillators emitting the black body radiation at a specific frequency overestimates this number and leads to diverging sums at high frequencies.

less energy. This miniscule amount of energy doesn't sound like much, but just remember that prolonged exposure to ultraviolet light quanta carrying 7×10^{-19} Joules can cause skin cancer.

By mathematically quantizing the number of photons available at each energy given by $E = h\nu$, Planck was able to uncover a single function that became the Wein approximation at short wavelengths and the Rayleigh–Jeans approximation at long wavelengths, which we now call the Planck equation (or function) for black bodies, written as:

$$B\,(\lambda,T) = \frac{2hc^2}{\lambda^5}\left(\frac{1}{e^{\frac{hc}{\lambda kT}}-1}\right)$$

The factor within the parenthesis in the formula arose because of the way that discrete quanta of light, each carrying an energy of $E = h\nu$ at a temperature of T, have to be counted. When the light wavelengths are very small, the exponential factor becomes very large and the resulting formula looks like Wein's Law. When the wavelengths are very large, the formula for $B(\lambda,T)$ looks like the Rayleigh–Jeans Law.

MAX PLANCK

Max Planck (1858–1947) went to school in Munich, where he became interested in mathematics and physics. He was also a proficient singer and composed music for piano and cello. In a 1924 lecture, Planck noted how he was advised while in college that it was pointless to study physics because 'almost everything had already been discovered.' By 1880, he had presented his PhD dissertation on the thermodynamics of bodies at different temperatures, and went on to investigate black body radiation in 1894. After several false starts he landed on the principle of quantization, although he was initially hostile to applying statistical principles. He did so out of despair since nothing else worked to derive the black body spectrum mathematically.

Planck's proposal was very troubling to many physicists because the work by Christiaan Huygens (1629–1695) in the 17th century and Maxwell in the 19th century meant that there was no dispute that light was incontrovertibly a wave phenomenon – not particulate. Even Planck wasn't 100 percent committed to the idea. He had invented the mathematical technique of quantization, yet he felt that this procedure was, in modern-day terms, something like buying a package of sausages or a six-pack of beer. There was nothing about either the meat or the ale that was inherently quantized, but both substances could be bartered in these convenient packages nonetheless. This is similar to Boltzmann's use of the atomic model for matter to further his mathematical goals in the study of heat and thermodynamics. Boltzmann, however, believed in the reality of these mathematical atoms, while Planck did not share such a conviction about light quanta.

THE PHOTOELECTRIC EFFECT AND LIGHT QUANTIZATION

The realization that light was quantized did not take hold until Albert Einstein (1879–1955) published his work in 1905 on the photoelectric effect. For Planck, quantization was a mathematical trick for deriving the correct form of the light energy distribution. For Einstein there was no trick involved; quantization was, instead, a fundamental phenomenon of Nature. How could it be that the same phenomenon, light, could appear as Maxwell's continuous electromagnetic waves under some conditions, and as Planck's particulate quantum under others? As quoted by historian Gerald Holton in his 1988 article *The Roots of Complementarity*, Einstein later reflected in 1924, 'We now have two theories of light, both indisputable, but it must be admitted, without any logical connection between them, despite twenty years of colossal effort by theoretical physicists.'

Independently of the work by Planck and the issue of the shape of Kirchhoff's universal spectral function for light, an entirely separate series of investigations had been in progress to understand how light interacted with matter. This research began with Edmond Becquerel (1820–1891) and his 1839 discovery of the photovoltaic effect, by which some materials increased their voltage when light was applied to them. This led to the

discovery by Willoughby Smith (1828–1891) in 1873 of photoconductivity in selenium, which led to the invention of the first photocells for generating electricity from sunlight.

In 1887, Heinrich Hertz (1857–1894) discovered that when his spark-gap transmitter of radio waves was placed in a dark box to better observe the spark, the gap had to be reduced in order to make the spark appear, but when exposed to ultraviolet

Heinrich Hertz's work on radio waves uncovered an effect that presaged the photoelectric effect.

light, the gap could be increased. The so-called Hertz effect was not pursued by Hertz himself, but it led to many investigations between the interaction of ultraviolet light with matter and the production of photoelectric effects. What was quickly discovered was that when light illuminated the surfaces of some metals, electrons would be ejected, and the energy of these electrons depended upon the frequency of the light but not its intensity. This was a result in contradiction to classical light theory, which would have predicted that the higher the light intensity striking a surface, the more energy would be delivered to the electrons and so the electron energy should be proportional to the intensity of the light.

There was also a thresholding effect for different materials. Light with a lower frequency would not eject any electrons from a surface, no matter what the light intensity. Above a specific frequency, however, electrons would be emitted and the number depended on the intensity of the light, but the energy of those ejected electrons above the threshold would

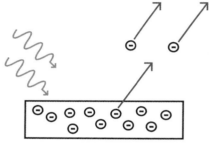

In the Photoelectric Effect, electrons are emitted when the energy carried by incident photons reaches a critical energy for the material, which is independent of the light intensity.

not change with light intensity. Only the light frequency above the threshold frequency was related to the electron energy. By 1904 these relationships were well measured, but there was no understanding of them using classical electromagnetic wave theory for light.

In 1905, Einstein borrowed Planck's light quantum idea and proposed that light energy was indeed quantized in terms of the formula $E = h\nu$, and that this real characteristic of light explained why the photoelectric effect behaved as it did. Every substance has its own threshold energy, W, and the electrons are trapped in the substance when their energies are below this critical energy. This translates by $E = h\nu$ into a critical frequency (wavelength) for light energy. When the individual photons interact with the electrons at energies below W, there is no photoelectric effect. But when the photons have a frequency above this critical value determined by $W = h\nu$, the electrons will be ejected with an energy equal to the difference ($E - W$). The number of electrons ejected depends on the number of photons available above the critical energy W, but this increase in photon intensity only changes the electron current and not the energy of the electrons themselves.

Einstein, who never received the Nobel Prize for his theory of relativity, did win the 1922 Nobel Prize for Physics and his work on the photoelectric effect. Einstein never liked the idea of nature being quantized, but it was his explanation for the photoelectric effect in matter, using Planck's light quantum idea as a real feature of nature, that pioneered the entire Quantum Revolution itself.

Albert Einstein was the first to explain the photoelectric effect as a consequence of Max Planck's quantization of light energy.

Key Points

- A black body is an object that emits electro-magnetic radiation (e.g. light) such that by Kirchhoff's Law, all of the radiation falling upon it is completely absorbed and re-emitted as a spectrum of light with a unique shape.

- Black body curves result when light is in thermal equilibrium with a body at a fixed temperature, and the shape of the emitted spectrum depends only upon the body's temperature.

- Planck explained this effect by proposing that light was itself quantized into discrete packets of energy, but he considered this a mathematical 'trick' to facilitate his calculations.

- Albert Einstein proved that light was quantized by explaining the photoelectric effect in which matter ejects electrons in a specific way.

- Wein's Displacement Law relates the temperature of a black body to the wavelength where the peak of its emission occurs.

- The Stefan–Boltzmann Law relates the emitted light power by a black body to its temperature and is an invaluable relationship for understanding the light emission and luminosity of stars.

CHAPTER 3
Matter Waves

In the previous chapters, we saw that Einstein proposed that light energy was actually quantized according to Planck's E = hν in 1905, a time when the most recent model for the atom was Sir Joseph John Thomson's Plum Pudding Model in which atoms consisted of a uniform mixture of positive and negative charges. This model was replaced by the Rutherford Model in 1911 with its heavy, positively charged nucleus surrounded by a cloud of electrons. However, once Bohr modified this view into a picture of just a few planet-like electrons for light atoms, the Rutherford–Bohr model caught the imagination of the public. It has since been continuously used as a symbol for atoms and even for 'atomic' energy (even though this is more properly considered nuclear energy).

The Rutherford–Bohr planetary model for the atom, and its upgrading of the contents of the nucleus from the discovery of the proton and neutron, is the simple model we all carry around in our heads and were taught at school. This model provided all of the right ingredients to finally explain the nature and chemical organization of the elements. Elements were chemically fixed because their atomic mass M, and atomic number Z, were dictated by the total number of positively charged nuclear protons (Z) and the sum of the nuclear protons and neutrally charged neutrons (N) such that $M = Z + N$. Radioactive decay, discovered by Henri Becquerel and studied in detail by Marie and Pierre Curie, could affect Z and M. An element that emitted alpha particles, which were nothing more than helium nuclei with Z = 2 and M = 4, would change into a lighter element

RUTHERFORD-BOHR MODEL

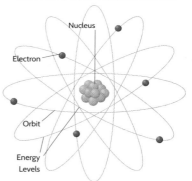

The Rutherford–Bohr atom also including the nuclear neutrons and protons.

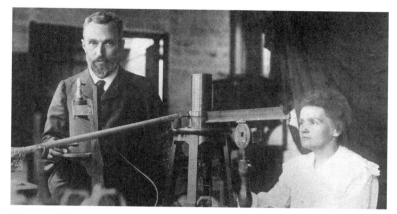

Pierre and Marie Curie discovered the properties of radioactive atoms.

with atomic number Z - 2, and an atomic mass of M - 4. For instance, Uranium (Z = 92) with a mass of M = 235 will emit an alpha particle and decay to Thorium (Z = 90) with a mass of M = 231. This decay takes about 700 million years. Purely chemical reactions, meanwhile, only affect the number and arrangement of the planetary electrons.

What was amazing was just how empty an atom seemed to be. Atoms are typically about 10^{-8} centimetres across – a unit of length called an angstrom. Nuclei, however, are one million times smaller and comparable to 10^{-14} centimetres – a unit of length called the fermi. If an atom were magnified to the size of a football, the nucleus would be smaller than a grain of sand at its centre, yet this nucleus would be the cause of nearly all the mass of an atom since electrons are 1/1836 times less massive than protons.

BOHR ORBITS AND MOMENTUM QUANTIZATION

Although intuitively elegant, the planetary electron proposal ran into severe problems. The significant electrostatic acceleration that the positive and negative charges would experience meant that the electrons would be undergoing huge amounts of acceleration. To see why this arises, consider that Coulomb's Law of the electrostatic force states that the force between two charges, Q and q, is given by

$$F = k \ \frac{Q \, q}{r^2}$$

where k is Coulomb's constant (8.99×10^9 Newton m^2/C^2), Q and q are the charges of the two particles in units of Coulombs (C) and r is their separation in metres. For the electron and proton in the hydrogen atom, r $=10^{-10}$ metres and $|q| = |Q| = 1.6 \times 10^{-19}$ C so the force between them is 2.3×10^{-8} Newtons. This means that the electron with a mass of 9.1×10^{-31} kg feels an acceleration of F/m $= 2.5 \times 10^{22}$ m/s^2. From simple calculations using Maxwell's electrodynamics, electrons undergoing this kind of acceleration should be emitting copious amounts of electromagnetic energy, and within 0.01 microseconds should all collapse into their atomic nuclei. In other words, the planetary model for atomic electrons was completely unstable.

A second problem was that this orbit decay radiation should follow a continuous emission spectrum with constantly increasing frequency – in line with Maxwell's electrodynamic theory, which stated that if the applied acceleration was uniform over time, there was no mechanism whereby the emission could proceed in fits and starts. That meant emission of light by the orbiting electrons could not reproduce the very narrow atomic spectral lines at specific frequencies that had been spectroscopically observed for nearly 100 years. In 1913, Bohr solved these two problems in a single masterful stroke by combining the Planck/Einstein light quantum idea with his

Niels Bohr developed a new theory for the atom that incorporated quantum effects.

planetary electron model, but adding a new, and deeply puzzling, law to atomic physics.

Bohr proposed that electrons orbited the nucleus in discrete orbits in an almost identical manner to the way that planets orbit our sun in specific paths. Within these orbits and no others, electrons miraculously did not radiate any electromagnetic energy at all despite being steadily accelerated by the intense nuclear electrostatic field. An electron in one of these stable orbits could, however, suddenly jump to another orbit representing another energy level by emitting or absorbing a quantum of light whose energy was precisely equal to the difference in energy between the initial and final orbits.

The location of each stable orbit was dictated by Bohr's condition that the angular momentum of an electron in each orbit would be quantized in units of L = nh, where n is an integer and h is Planck's constant. By comparing the electron orbits to the

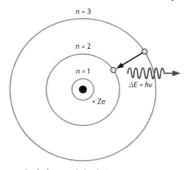

Bohr's model of the atom.

properties of planets according to Kepler's Third Law such that the square of the period of the orbit, T^2, is proportional to the cube of the distance from the nucleus, a^3, Bohr created a self-consistent model for the hydrogen atom that led to the formula derived by Balmer for the spectral line emission. In fact, the result was a complete re-derivation of Balmer's formula, but where the constant of proportionality – Rydberg's Constant, Re – could now be directly calculated from other physical quantities rather than having to be experimentally deduced. The Rydberg Constant is now defined as

$$Re = \frac{Z^2\,(ke^2)m}{h^2}$$

for which k is the Coulomb constant, e is the charge of an electron,

m is the mass of an electron, Z is the atomic number of the atom, and h is Planck's Constant. With appropriate values, and for the hydrogen atom with $Z = 1$, $Re = 2.2 \; 10^{-18}$ Joules. Because these quantities of energy are so small, physicists use another unit of energy called the electron-volt (eV) such that 1 eV = 2.2×10^{-18} Joules, which results in Re = 13.6 eV. If a single photon carried exactly this much energy, it would completely eject the electron from the hydrogen atom, causing the previously neutral atom to become fully ionized. A photon of that energy, according to E = hv, would have a wavelength of 91.2 nm – the Lyman limit. Photons at shorter wavelengths lack the energy to ionize a

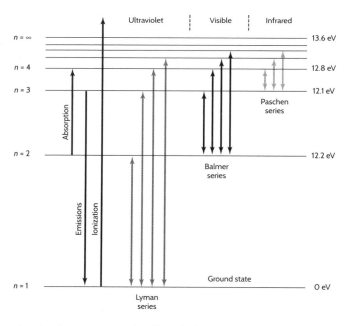

Figure showing the main energy levels in a hydrogen atom and the relationship between the spectral line sequences and the energy differences. For instance, the first line in the Balmer series, called Balmer-alpha (H-α) is produced when an electron jumps from an energy level of 12.1 eV to 10.2 eV. This energy difference of 1.9 eV, according to the quantization condition E = hv produces a spectral line with a frequency of $v = 4.57 \times 10^{14}$ Hz and a wavelength of $\lambda = 656.45$ nm.

hydrogen atom, but photons in the ultraviolet spectrum at wavelengths shorter than the Lyman limit can fully ionize hydrogen atoms, creating a plasma state. These are commonly found in astronomical systems called emission nebulae, such as the Great Nebula in Orion.

The Bohr model worked very well for hydrogen, but it became prohibitively complicated and inaccurate for multi-electron atoms for which the electrons interacted among themselves and caused complex orbits that could not be so easily quantized. However, it did manage to reproduce many basic features of the spectra of heavier, hydrogen-like atoms with the additional idea that these orbital shells could be populated by more than one electron, and that electrons in the same shell did not interact. Also, different shells produced screening of the central positive charges for shells further out from the nucleus.

NEILS BOHR

In 1905, while an undergraduate, Neils Bohr (1885–1962) won a national competition to measure the surface tension of liquids, and went on to investigate the electron theory of metals for his doctoral thesis in 1911. Because it was written in Danish, this ground-breaking work never received much attention. Travelling to England, where most of the work on atomic structure was being conducted, he joined Ernest Rutherford and went on to combine Rutherford's heavy nucleus model with Planck's quantization idea and created a model that accounted for the spectral line pattern found in hydrogen atoms. Later, in 1919, he abandoned this, and went on to develop the Correspondence Principle. In 1922, he was awarded the Nobel Prize in physics.

Despite its imprecise predictions for multi-electron atoms, the Rutherford–Bohr model held considerable sway among 20th-century chemists and was a useful heuristic model for physicists and educators for decades. Even today, it is still the preferred first atomic model encountered by school pupils, with the addition of the discovery of the proton and neutron to flesh out the nuclear details. However, the most mysterious ingredient to this model was to be found in the behaviour of the electrons themselves.

ELECTRONIC STRUCTURE

In an October 1939 BBC broadcast, Winston Churchill once remarked about Russia that it was 'a riddle wrapped in a mystery inside an enigma'. No less a description is true of the electron, cloaked as it is in its electromagnetic field and stuffed inside the equally mysterious atom. But in the early 20th century it would soon be recognized that there were even stranger aspects to the study of this fundamental kernel of matter.

Experiments by Walter Kaufmann (1871–1947) and Alfred Bucherer (1863–1927) in 1901 led to the surprising result that the increase in the mass of a charged particle as its speed increases was identical to what had been predicted for its electromagnetic mass alone from Maxwell's Theory of Electrodynamics developed in the mid 1800s. What it implied was that the mass of an electron, which modern tables list as 9.11×10^{-31} kilograms, is *completely* defined by its electromagnetic field. There is nothing left over in the experiments, no solid clump of 'Newtonian' matter buried at the heart of an electron, to give it its measured mass.

If the mass of the electron was somehow a product of its electromagnetic field alone, what was the correct way to think about its structure? It was fashionable in 1904 to think of the electron as a small sphere carrying its charge on its surface, like a planet covered with an ocean. But what was holding it together against the enormous forces of electrostatic repulsion that must surely be trying to rip it apart? The French physicist and relativist Henri Poincaré (1854–1912) offered a way out of this problem by suggesting that compensating forces within the electron would hold it together and cause it to be perfectly rigid as it moved. This idea turned out to be flawed upon closer mathematical analysis. Such a rigid object with a finite size could never be consistent with the equally rigid predictions offered by Einstein's theory of special relativity. These kinds of objects would violate the basic principle of relativistic invariance, which stated that no experiment could reveal the absolute motion of an observer. The shape of the electron would

Henri Poincaré proposed a model for the structure of the electron as a rigid sphere.

be different as judged by observers each travelling along different paths in different reference frames, which violates Einstein's relativity principle.

In Germany, Wolfgang Pauli (1900–1958) believed that the electron's structure was one of the most pressing issues in physics, and he devoted many of his early years in physics to studying it. Pauli eventually concluded that the space inside an electron may in some way be physically different than the space outside it. A natural dividing line would be what was called the Compton radius of the electron, where the electrostatic potential energy of the electric field equalled the rest mass of the electron; a distance of 2.8×10^{-15} metres. This didn't sit too well with Albert Einstein, whose theories of special and general relativity required that space, and in fact spacetime, be continuous and indivisible no matter where they were located. Moreover, Einstein's equations for high-speed particles showed that they would gain mass in a specific way. The measurements of the growing mass of an electron with speed were entirely consistent with the purely electromagnetic and absolutely point-like nature of the electron's mass.

So, by the early years of the 20th century, the simplest picture of the electron had to show it as a vanishingly small particle, perhaps even a mathematical point with zero physical dimension. There would be no solid surface to it, just a thickening knot of electromagnetic energy anchored to a specific point in space. Yet this knot of energy possessed mass and charge, although what these words might actually mean for such a point-like particle, no one quite knew. Outside the atom, electrons could create continuous electrical currents that produced magnetic fields, but inside the atom, electrons were in some way the midwives to the emission of discrete spectral lines.

MATTER WAVES

By the time the Rutherford–Bohr atom was making its debut, light was regarded as having both wave-like and particle-like properties, which allowed both electromagnetic calculations to be performed in the wave-like picture, and quantum calculation to be performed for black body and spectral line emission in the particle-like model. This wave-particle duality was, itself, enormously frustrating to dwell upon because it was quite clear under strict laboratory conditions that the same phenomenon (light) yielded its particle aspects when one class of physical problems involving the production of light was considered, and definite wave-like aspects such as interference and diffraction when another class of problems were considered. As Einstein wrote in 1938, 'It seems as though we must use sometimes the one theory and sometimes the other, while at times we may use either. We are faced with a new kind of difficulty. We have two contradictory pictures of reality; separately neither of them fully explains the phenomena of light, but together they do.' What was even stranger is that matter did not seem to partake of a similar wave-particle duality – or did it?

Experiments had shown rather consistently that matter seemed to be thoroughly particle-like, so the symmetry between matter and light seemed to be broken. As we saw in the previous section, the nature of the electron as a purely electromagnetic entity seemed to demand a wave-like description consistent with its electrodynamics. Louis de Broglie (1892–1987) decided to explore the issue of whether matter also had a dual nature in his 1924 PhD thesis, and uncovered a surprising result: 'When I conceived the first basic ideas of wave mechanics in 1923–24, I was guided by the aim to perform a real physical synthesis, valid for all particles, of the coexistence of the

Louis de Broglie proposed that matter had wave-like properties, which led to a revolution in quantum mechanics and atomic theory.

wave and of the corpuscular aspects that Einstein had introduced for photons in his theory of light quanta in 1905.'

Through a simple mathematical manipulation of two key formulae, de Broglie discovered that the momentum of a particle, $p = mv$, and its Planck energy $E = hc/\lambda$, could be rewritten with a bit of algebra so that $E = mc^2$ becomes $E = p/c$, so that $p/c = hc/\lambda$. This leads directly to $\lambda = h/p$. A massive particle with a momentum of p could be identified as having a wavelength λ. For example, an electron travelling between the plates of an electrostatic capacitor with a voltage difference of 100 Volts reaches an energy of 1.6×10^{-17} Joules, and has a speed of 6×10^{6} m/s. For a mass of 9.1×10^{-31} kg, this results in a momentum of $p = 5.4 \times 10^{-24}$ kg m/s. Using de Broglie's matter wave formula with Planck's constant $h = 6.6 \times 10^{-34}$ m^{2}kg/s, we get a wavelength of $\lambda = 1.2 \times 10^{-10}$ metres. There is apparently no obvious limit to this matter

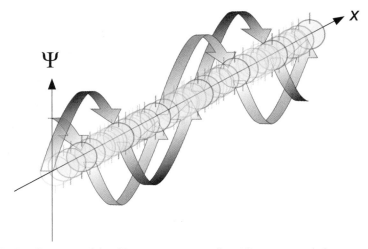

Ψ

X

De Broglie proposed that his waves were actually guiding matter to behave as wave-like objects, but were not themselves waves of matter.

wave effect, which can even be applied to macroscopic objects such as humans ($\lambda = 5 \times 10^{-36}$ metres), although to what purpose is a matter for the imagination.

In 1927, a few years after de Broglie's discovery, Clinton Davisson (1881–1958) and Lester Germer (1896–1971) fired slow-moving electrons at a crystalline nickel target and found that the electrons displayed the same diffraction pattern as those for X-rays. Matter did in fact display unambiguous wave-like characteristics. In 1999, a research team in Vienna demonstrated diffraction for molecules as large as carbon-based fullerenes such as C_{60} with an inferred wavelength of 2.5 picometres. More recent experiments involving molecules made of 810 atoms with a mass of 10,123 amu have also confirmed this matter wave effect, so matter does indeed have wave-like attributes.

The earliest application of de Broglie matter waves was in its explanation of quantized orbits in the atom. Each orbit was now defined in terms of an electron matter wave having an integer number of wavelengths around the circumference of the orbit. For other locations this would not be the case and the electron's

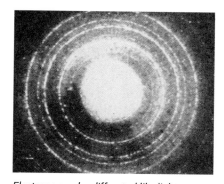

Electrons can be diffracted like light waves, in this case in a crystal of aluminium oxide.

matter wave would interfere with itself and not be possible. This led Bohr to propose an improved model for the atom, which could now naturally explain electron orbitals, and in particular explain why angular momentum had to be quantized as he had proposed in the earlier 1913 Bohr atomic model.

The earliest application of de Broglie matter waves was in its explanation of quantized orbits in the atom. Each orbit was now defined in terms of an electron matter wave having an integer number of wavelengths around the circumference of the orbit. For other locations this would not be the case and the electron's matter wave would interfere with itself and not be possible. This led Bohr to propose an improved model for the atom, which could now naturally explain electron orbitals, and in particular explain why angular momentum had to be quantized as he had proposed in the earlier 1913 Bohr atomic model.

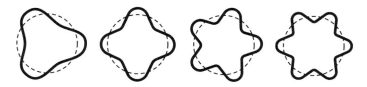

De Broglie electron waves cause the orbits of electrons in atoms to be found only in certain locations where the waves do not interfere with themselves.

De Broglie's matter waves led to the quantization condition that $2\pi r = n\lambda$, where $n = 1, 2, 3...$ But since from de Broglie $\lambda = h/p$, we have $2\pi r = nh/p$ and therefore $mvr = nh/2\pi$, which shows how angular momentum is quantized.

Key Points

- The Bohr Model introduced the idea of momentum quantization to explain why electron orbits came only in specific energy intervals.

- Bohr's Model explained for the first time why only specific spectral lines were seen in the spectrum of hydrogen, but could not explain lines from more complex atoms.

- De Broglie's discovery of matter waves allowed both light and matter to share a dual wave-particle nature, which led to an explanation for Bohr's orbit quantization.

- The central puzzle of the structure of an electron is that it must be a point-like particle defined entirely by its electromagnetic properties with no hidden mass, while at the same time behaving as both a wave and a particle.

- The Davisson and Germer experiments on electron diffraction proved that electrons had wave-like characteristics, and by extension, all forms of matter had this quantum property, even human beings.

CHAPTER 4
Wave Functions and Quantum States

The realization that matter had wave-like attributes led to a technical improvement of the Rutherford–Bohr atom, but this did not cure all of its problems in dealing with multi-electron atoms. There was some deeper aspect to matter waves and light quantization that still needed to be ferreted out from this pastiche of empirical and heuristic theories, and developed into a single, detailed theory akin to Newton's gravitational mechanics, Maxwell's electrodynamics and Einstein's relativity. It is here that we encounter one of the most revolutionary ideas in nearly all of physical science, and arguably one of the most difficult ones to describe accurately to the non-scientist without the use of detailed mathematical exposition.

One of the first to make the application of matter waves to atomic structure more mathematically rigorous was the Austrian physicist Erwin Schrödinger (1887–1961). His insight, published in 1926, was an elegant equation describing how a matter wave should evolve in time and space. His equation, which we will describe below, used a well-known mathematical technique to describe waves, but in order to apply this to sub-atomic objects such as electrons, he had to come up with a suitable mathematical object that would be 'waving' in the first place. Nothing succeeds like success and in the end, Schrödinger successfully used his wave equation to derive the energy quantum levels of hydrogen in a more rigorous way than Bohr. However, there was a twist in the interpretation of what exactly was waving. Most physical waves can be interpreted as a change in some physical quantity such as the height of a surface (water waves), the ambient pressure (sound waves), space curvature (gravity waves) or the strength of an electromagnetic field (light wave). For electron waves, in order to make the mathematics work, the object that was waving was not a physical medium, but a mathematical object whose complex amplitude-squared ($\Psi \times \Psi^*$) represented the probability of an electron being in a specific place at a particular time according to an interpretation by Max Born in 1926. The object was called a wave function and represented by the Greek symbol Ψ.

Schrödinger's matter wave equation was intended to be the analogue of Maxwell's equations for electromagnetism and light and takes the following form:

$$i\hbar \frac{\partial \Psi}{\partial t} = - \frac{\hbar^2}{2m} \nabla^2 \Psi + V\Psi$$

This looks formidable, even cryptic, but its overall form is that of the ordinary Newtonian equation for the total energy, E, of a particle

$$E = \frac{1}{2} mv^2 + V(r)$$

that consists of the sum of its kinetic energy [$1/2\ mv^2$] due to its motion, and its potential energy [$V(r)$] due to its location within a field of force. In this instance for electrons orbiting an atomic nucleus, m is the mass of the electron and $V(r)$ is the function that gives the Coulomb electrostatic potential energy of the particle with a charge of e, orbiting a nucleus with a charge Ze, at a distance of r, given by

$$V(r) = k\ \frac{Ze^2}{r}$$

The way to use Schrödinger's equation is first to define the potential energy function, $V(r)$, for the system and then to solve the equation for the total wave function Ψ. This mathematical function will have a form suitable for describing the particular system under study but there is a catch. As we will see later, the total wave function actually consists of an infinite family of functions – $\varphi_{n,l,m}(x,t)$ – one for each possible state of the particle, indexed by their quantum numbers n, l and m, where n is the quantum number for the energy level, l is the quantum number for orbital angular momentum, and m is the so-called magnetic quantum number that identifies which suborbital an electron is in. The particular wave function you are looking for is the one whose quantum numbers match the ones you are interested in, or have experimentally measured. Once you have selected the quantum numbers of the particular state you are interested in, you can calculate the probability that the electron is located somewhere in space (x) and time (t) within the quantum system by simply plotting out what the function $\varphi(x,t) \times \varphi(x,t)^*$ looks like for its specific quantum state, where $\varphi(x,t)^*$ is the complex conjugate of $\varphi(x,t)$.

ERWIN SCHRÖDINGER

Erwin Schrödinger (1887–1961) completed his doctorate in 1914, and his early work in physics was in the fields of electrical engineering and atmospheric electricity. He studied the then new quantum ideas by Max Planck and Albert Einstein, and by 1922 he had become interested in Herman Weyl's ideas of geometry applied to electron orbits in the atom. This led to his 1926 paper 'Quantization as an Eigenvalue Problem', which was hailed as one of the most important papers in 20th-century physics. In it he developed a simple equation that now bears his name, and a detailed mathematical technique which is simple to apply to all quantum systems, involving his new idea of the 'wave function' to describe quantum states. Philosophically he never liked this perspective, and once said that he had been sorry he had ever had anything to do with its invention and subsequent development.

Erwin Schrödinger.

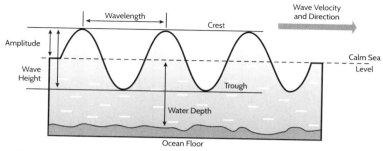

Ordinary water waves are defined by the vertical displacement of the height of the water surface. The mathematical equation that describes the change in time of the water height is called a wave equation.

You may have noticed that the term on the left-side of the equals sign in Schrödinger's equation involves an imaginary number represented by the symbol i, which is shorthand for $i= \sqrt{-1}$. This was the genius of Schrödinger, to recognize that the Newtonian equation for energy could be retranslated into an equation for wave functions, but only if those wave functions were complex numbers such as A + Bi. The mathematics of complex algebra has to be used to describe Ψ. A complex number can be converted into a real number like probability, by multiplying itself by what is called its complex conjugate. For example, the complex conjugate of A + Bi, written as (A + Bi)*, is A - Bi so that the product of the number with its complex conjugate is (A + Bi)(A - Bi) = A^2 + ABi − BAi + (+Bi)(-Bi) = A^2 + B^2, which is a real number. So Schrödinger's wave function was related to the probability P= Ψ x Ψ* which was a real number between 0 and 1. In a sense, the amplitude of the wave function Ψ defined in Schrödinger's mathematics was the square-root of the probability of finding an electron at a particular place, x, and time, t, inside the atom. In fact, what was actually intended by the wave function was technically far more complicated.

The wave function represented the specification for an electron in one quantum state within the atom, which was indexed by the quantum number for its energy, n, its angular momentum, l, and the projection of its angular momentum along the coordinate z-axis, *m*, which could be determined by applying a magnetic field to the atom. So the correct way to describe the state of the electron was through a function, $\varphi_{n,l,m}$ (x,t), where the variable *x* is a placeholder for all three spatial dimensions such as (x,y,z) in Cartesian coordinates or (r, θ, φ) in spherical coordinates. Taking advantage of the spherical symmetry of an atom, this function φ can actually be split into the product of two functions called a spherical Legendre polynomial, $Y^m_l(\theta,\phi)$, and a radial Hermite polynomial, $H_n(r)$, so that $\varphi_{n,l,m}(r,\theta,\phi,t) = H_n(r)\ Y^m_l(\theta,\phi)$.

In terms of Schrödinger's wave function, you take the wave function for a particle in a specific state and compute

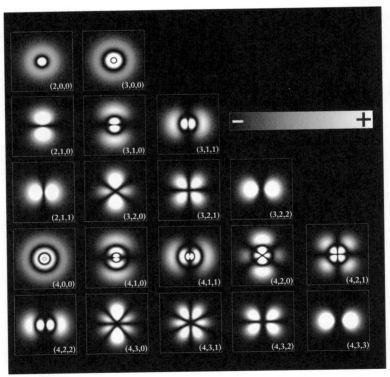

A few of the computed elementary electron probability clouds P= $\phi_{n,l,m}(x,t)^*$ $\phi_{n,l,m}(x,t)$ for the hydrogen atom, along with their quantum numbers.

the probability function $P(n,l,m,x,t) = \varphi_{n,l,m}(x,t)^* \; \varphi_{n,l,m}(x,t)$. Mathematically, this quantum state function is just one element of the full wave function Ψ of the electron. The Hungarian mathematician John von Neumann (1903–1957) explained in his masterful book *The Mathematical Foundations of Quantum Mechanics* (1932) what he had first showed in 1926: that each of these individual quantum states is collected together into what is called a Hilbert space, for which each point is one possibility for the state of the system defined by $\varphi_{n,l,m}(x,t)$.

HEISENBERG'S UNCERTAINTY PRINCIPLE

Between 1923 and 1925, Werner Heisenberg (1901–1976) investigated another way to represent quantum systems other than via Schrödinger's quantum theory involving wave functions. His approach was that only measurable features of an atom can be a part of the mathematical treatment of the state of the atom. Between measurements, an electron in an atom was not a defined physical object. It came into existence only at the moment of observation so that instead of a set of probability waves, an electron in a quantum state was defined as a matrix of values that represented the many observable outcomes of an observation. The two fundamental matrices were those of position Q and momentum P, and these matrices obeyed a relationship called non-commutation. Matrices are said to commute when multiplied together if $A \times B = B \times A$, but for Heisenberg's position and momentum matrices they followed the non-commuting relationship for which $AB \neq BA$ and this led to Heisenberg's matrix quantum condition $[PQ - QP] = ih/2\pi$. By 1925 Heisenberg had developed a scheme for calculating the spectral line frequencies for hydrogen. The calculations were generally considered to be tedious and rather cryptic, so few physicists were willing to follow his laborious discussions fully, but within his matrix mathematics were hidden some crucial insights about how quantum systems are measured.

WERNER HEISENBERG

Werner Heisenberg (1901–1976) received his doctorate in physics in 1923 under the mentorship of physicist Max Born and Arnold Sommerfeld, and the mathematician David Hilbert. In 1923 he began work with Niels Bohr at the University of Copenhagen, where he collaborated with Hans Kramers on the mathematical description of the scattering of light by atoms, and developed his matrix approach to calculating how atoms interact with light. His groundbreaking but mathematically difficult 1925 paper introduced physicists to matrix mathematics, a tool never before used to describe physical systems. He was later awarded the 1932 Nobel Prize in Physics, having been nominated for the prize by Albert Einstein.

Werner Heisenberg.

The first insight was that, before measurement, the state of a system is one among a vast collection of possible outcomes represented by the enormous collection of elements in the Heisenberg matrices. So it is not meaningful to talk about what an electron is 'doing' between measurements. The analogy devised by Heisenberg was to be sitting in a park at night watching a person travel from one lamp post to another. You cannot see what the person is doing in the dark, only how they suddenly appeared at each lamp post. For the electron, this almost literally means in the Heisenberg interpretation that *electrons do not really exist* between measurements, and Heisenberg's matrices list all of the possible outcomes of the observation or measurement of what the electron is doing when it arrives at the 'lamp post' and is physically measured. This viewpoint was carried over to Schrödinger's quantum theory by saying that before measurement an electron is manifesting all possible measurement outcomes specified by its total wave function, Ψ, but that after measurement, all of these possibilities suddenly collapse into only one state with specific, measurable quantum numbers, $\varphi_{n,l,m}(x,t)$. This became known as the Copenhagen Interpretation of wave function collapse, and remains a difficult issue in quantum theory today.

The issue of whether the wave function of a particle actually collapses or not has been challenged by such ideas as the many-worlds interpretation offered in 1957 by Hugh Everett

Measurement of an electron is like watching a man walking between lampposts on a dark night. He suddenly appears at each lamp post, and seems not to exist in between.

(1930–1982) at Princeton University. Everett was a student of John Archibald Wheeler, and Everett's solution for avoiding wave function collapse, offered in his PhD thesis, was met with considerable scorn. He left physics immediately after receiving his PhD, although by the 1970s his idea had become very popular among physicists working in this area of research. According to the many-worlds interpretation, the wave function of a system, Ψ, no matter how complex, never collapses for observers within our universe. Instead, with each quantum measurement or observation, a new universe splits off from the previous universe. This process of splitting allows the wave function across all of the spawned universes to remain un-collapsed while observers within each separate universe evolve along the many pathways made possible by each measurement outcome defined by the specific quantum state measured, $\varphi_{n,l,m}(x,t)$.

Heisenberg's second insight is a bit more difficult to motivate. The momentum and position of an electron can be represented as complimentary components, just as frequency and wavelength are for describing light waves. In the language of a mathematical

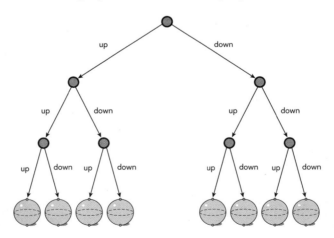

A figure showing how the universe splits into multiple copies, depending on whether a single photon is measured to have a spin up or a spin down after three successive measurements.

framework called Fourier analysis, quantities such as position and momentum, or energy and time, can be represented as Fourier pairs described in wave-like terms. This means that if you know one of the pair members with great precision, the complimentary pair member cannot be known with equal precision. For example, this figure shows an electrical signal that is localized to a specific time, called an impulse, which exists

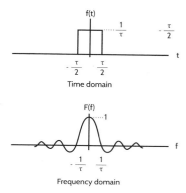

Time domain

Frequency domain

Although an electrical impulse is confined to a small time interval in the Time Domain in the Frequency Domain it is spread out.

only between a time -τ/2 and +τ/2, but its Fourier partner in the frequency (energy) domain is a complex wave train that spreads out. Similarly, when you try to localize a particle to a specific point in space, its momentum description spreads out and cannot be simultaneously specified with precision. In other words, the more you try to localize where a particle is in space, the worse becomes your knowledge about its exact momentum or speed.

This reciprocal behaviour between momentum, p, and position, x, and also between energy, E, and time, t, is the core of Heisenberg's Uncertainty Principle, which was introduced by Heisenberg in 1927, and derived by Earle Kennard (1885–1968) and Hermann Weyl (1885–1955) in 1928. Heisenberg's Uncertainty Principle is generally stated as a pair of equations that relate the uncertainties, Δ, in the x and p, and t and E components as

$$\Delta_x \Delta_p \geq \frac{h}{4\pi}$$

$$\Delta_t \Delta_E \geq \frac{h}{4\pi}$$

At the atomic scale, if you know the location of an electron to within 1 nanometre ($\Delta_x = 1$ nm) then you cannot know the speed of this particle to better than 58 km/s [e.g. $\Delta p \geq h/(4\pi\Delta x)$]. The application of Heisenberg's Uncertainty Principle to uncertainties in energy will be discussed in Chapter 9.

So, by 1927, virtually all of the ingredients to a self-consistent quantum theory of the atom had been mathematically established, and the combination of Schrödinger's elegant and intuitive mathematics together with Heisenberg's uncertainty relations quickly led to accurate descriptions of atomic structure and predictions for measurement outcomes. However, there were still several problems outstanding. First, Schrödinger's quantum theory was not as yet consistent with special relativity. Second, there were a whole host of issues related to the exact process of measurement and how the wave functions for quantum systems collapsed, which we will cover in the next two chapters.

Key Points

- Erwin Schrödinger developed a mathematics for describing the wave-like properties of electrons within atoms involving wave functions, which describe the probability of finding a particle in a specific quantum state.

- Werner Heisenberg developed an independent mathematical theory that ignored knowledge about where a particle was before the moment of its interaction with other quantum systems.

- Upon making an observation of the state of a particle, the full wave function of all possible states collapses into only one final state, which is called the Copenhagen Interpretation.

- Heisenberg's Uncertainty Principle sets a fundamental limit to how well two quantities such as energy and time, or momentum and location, can simultaneously be known with infinite precision.

- The Schrödinger and Heisenberg quantum theories led to the first precise and mathematically consistent explanation for the structure of atoms, from which detailed calculations could be performed.

CHAPTER 5
Quantum Measurement

One of the most puzzling issues that has arisen in the quantum description of matter is the issue of measurement and observation. For macroscopic systems, it is obvious that observation has no impact: when we observe a rock on a beach, the act of observation does not suddenly change the rock into something else. When we measure the intensity of sunlight with a camera, the measurement does not change either the sun or the object whose surface reflects the light. In the quantum world, however, this is apparently not the case. Moreover, the object being observed or measured does not apparently have a definite physical state or set of properties before the act of measurement is completed.

We have seen how an electron in an atom is represented by a generalized quantum state, Ψ, which gives all possible states of energy the momentum that it could assume. This is usually expressed by the following symbolism:

$$\Psi = \sum_{i=1}^{\infty} a_i \varphi_i$$

As we saw in Chapter 4, the Greek letter Ψ represents the complete wave function of the particle while φ_i represents the individual quantum states that the particle can assume based on specific quantum numbers for energy (n), angular momentum (l), and magnetic quantum number (m). The quantity a_i is a complex number related to the probability, P_i, of finding the particle in a particular state according to $P_i = a_i^* \, a_i$ where a^* is called the complex conjugate of the amplitude of the state.

QUANTUM SUPERPOSITION

Before a measurement or observation is made, the only information we have about it is its complete wave function, Ψ , which is a solution to Schrödinger's equation for the specifics of the system the particle is in. However, after the measurement, the particle appears in a specific state φ and *all* of the other possible quantum states have vanished. This is called the collapse of the wave function according to the Copenhagen Interpretation. The way in which this happens has been a deep theoretical issue since the 1930s. Another way to state this is to say that initially the particle is in a superposition of many quantum states such that Ψ $= a_1 \varphi_1 + a_2 \varphi_2 + a_3 \varphi_3$ + etc, but after the observation the particle is found in a specific state, say φ_3 with a probability of $a_3^* a_3$. The full state, Ψ, contains far more information about the particle than what is found in any particular measured state, so where does all of this information go? Does it get destroyed? Erwin Schrödinger proposed a thought experiment that dramatically described this conundrum.

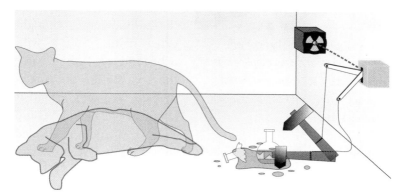

The Schrödinger's Cat experiment involves a cat placed in a sealed box with a radioactive source that randomly releases a poison gas. The cat is eventually placed in a mixed state of being alive or dead.

Place a cat in a closed box with a radioactive source that can, after a time, trigger a cyanide gas canister to kill the cat. After a while, there are now two possible states for the cat: alive or dead, so its state vector is $\Psi = \frac{1}{\sqrt{2}}Alive + \frac{1}{\sqrt{2}}Dead$. Note that the probability to be found in each of these two states is just $P = \frac{1}{\sqrt{2}} \times \frac{1}{\sqrt{2}} = \frac{1}{2}$. This is the conundrum of Schrödinger's Cat, which says that you can place even a macroscopic object in a quantum state of superposition for an indefinite length of time until you open the box and

Eugene Wigner was a physicist who proposed an even more complicated Schrödinger's Cat experiment to test whether the real world can be a mixed quantum state.

observe/measure that the cat is definitely either dead or alive. The question is, what physical act caused the cat's wave function to collapse? In a slightly more complex experiment called Wigner's Friend, we can place the cat and the observation in an even more contradictory condition.

In 1961, Eugene Wigner (1902–1995) considered the following thought experiment. Start with Schrödinger's cat inside the box. Before the box is opened, Wigner's Friend is about to open the box, and Wigner himself is located some distance away but cannot see into the box. We now have three 'particles' involved in this experiment: the cat, Wigner, and Wigner's Friend. When viewed by Wigner, Wigner's Friend opens the box and immediately sees which state the cat is in, but Wigner does not. He sees Wigner's Friend, who now knows the cat's state, but Wigner now finds Wigner's Friend in one of two states: either the state where the Friend saw a live cat, or a state where the Friend saw a dead cat. Now after a long period of time, Wigner continues to see the cat, the laboratory and Wigner's Friend in a superposed quantum state. So a few days, weeks or years later he finally asks his friend, Was the cat dead or alive? Only when Wigner's Friend finally tells Wigner the outcome, does the quantum state of the system collapse – but a portion of the quantum state had already collapsed for Wigner's Friend after he opened the box. The conundrum is that, depending on who the observer is, there is either a system in a definite state for Wigner's Friend, or a quantum-superimposed state for Wigner. Unless Wigner is accorded the status of a Privileged Observer, his friend's observation must also be credited as being valid, and this is where the quantum observation paradox seems to arise.

The Wigner's Friend experiment shows that two observers may not share the same objective reality, which refutes the idea that there is a single physical world that we all experience the same way.

Quite literally, Wigner and his friend are experiencing two very different realities within the same universe.

In 2019, Massimiliano Proietti at Heriot-Watt University in Edinburgh used advanced laser technology to reproduce the Wigner's Friend experiment by using photons that could have one of two possible polarization states. The experiment used six photons in a state of superposition (called quantum entanglement) so that one set of three photons represented Wigner and the other set of three represented Wigner's Friend. The second set recorded the polarization of a photon while also recording the state of the first set and whether it was still in superposition with the state of the photon. The result was that the Wigner set did indeed still record the Wigner's Friend set in a superimposed state even though it had measured and recorded the state of the photons. Both realities can coexist at the same time, so the observation paradox is a real feature of our physical world. This leads to a variety of important considerations, because we have always believed that science is based upon a consistent background reality, about which all observers will agree. The Wigner's Friend paradox says that under some situations there is no single underlying reality.

EUGENE WIGNER

Eugene Wigner (1902–1995) received his doctorate in science c. 1924 and temporarily went to work at his father's tannery before receiving a position at the Kaiser Wilhelm Institute in 1926, where he worked on x-ray crystallography. There he studied the quantum theory of Schrödinger and a mathematical branch called group theory, which pertains to the mathematical statement of symmetry. His articles on group theory and symmetry in atomic physics, published between 1928 and 1931, introduced physicists to symmetry principles, and led to an offer of permanent employment at Princeton University by 1937, when he became a naturalized US citizen.

LOCAL REALISM

Another issue in quantum mechanics is that a particle can behave either as a wave or as a particle but not both at the same time. For a given experiment, how does a particle 'decide' what to be? One experiment may be set up to measure the de Broglie wavelength of an electron, while another is set up to measure only its property

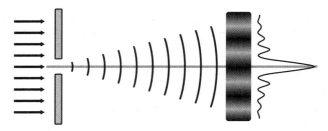

A single-slit experiment, where a light wave enters the slit on the left, and leaves behind a simple diffraction pattern of bright and dark lines on a projection screen.

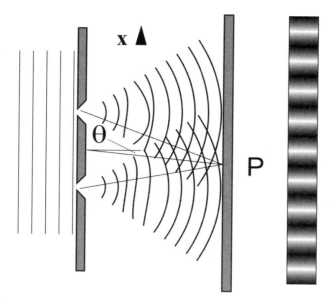

In a double-slit experiment, a light wave entering the two slits interferes with itself in the middle-region and leaves a distinct interference pattern on the screen.

as a particle but not a wave. How does the electron react? The classical experiment that reveals this dilemma is the double-slit experiment.

If light were purely a particle, it would still produce a diffraction pattern in the first experiment, but would not interfere with itself in the double-slit experiment. Instead, it would again leave behind a simple diffraction pattern.

In 1978 John Archibald Wheeler (1911–2008) proposed a delayed-choice experiment based on the ability of light waves to interfere with themselves in a double-slit experiment. A photon is emitted by a laser and passes through two slits, which produces a familiar interference pattern caused by the wave-like behaviour of light. But one of the slits is blocked, so the photon acts as a particle and passes through the open slit to produce a bright point on the screen as a particle should. How did the photon know at the time of creation that the slit was open or closed, so that it could register as either a wave or a particle on the screen? One possibility was that it reached the slit and then sent information back to the source to alter its properties accordingly. Because the travel times at the speed of light for laboratory equipment were nanoseconds, it is conceivable that this information could be transmitted quickly and still allow the photon to continue its smooth journey through the slit/s.

In 2015, Professor Andrew Truscott from the Australian National University Research School of Physics and Engineering used helium atoms at a temperature of nearly Absolute Zero to test Wheeler's delayed choice experiment. The procedure was similar to a double-slit experiment, but after the photon was sent through a laser beam splitter, a random number generator decided if the photon was to take a direct path to the detector or to interfere with itself and travel the longer path through the system. The result was that the atom behaved as though it did not 'make up its mind' until the instant of detection at the screen as to whether it would behave as a wave or a particle. Similar kinds of results have been obtained, most recently in 2017, when a satellite laser system was used to create two different paths for a particle or a wave: again the photon behaved simultaneously as both until the moment when the detector at the end of the line in the last instant framed the measurement one way or the other. In other words, it is the act of the measurement itself and the 'question' the experiment asks that determines whether a particle behaves as a wave or a particle. This takes us to a related issue commonly framed as Local Realism.

Local Realism states that a particle has a specific property independent of an observer measuring it. This is a view shared by both Sir Isaac Newton and Einstein. It is related to the philosophical idea that there is an objective reality that exists independently of whether an observer is present or not. The moon continues to exist even if every life form on Earth suddenly goes blind. But quantum mechanics says objects do not have a definable state until they are measured or observed. This is verified by both the delayed choice and the Wigner's Friend experiments, which describe macroscopic objects in a mixed or superimposed state that 'collapses' only after an observation is made. So quantum mechanics violates Local Realism.

Local Realism can be reintroduced into quantum mechanics if there are hidden variables present that turn quantum mechanics into a deterministic theory. De Broglie proposed that matter waves and particles were not incompatible descriptions because the matter waves, which he called pilot waves, were the signature of an underlying theory whose details we have yet to discover and that provide the missing parameter (variable) to make the description fully deterministic. To see how this works we have to use particles in an entangled (superimposed) quantum state.

In 1935, Einstein, Nathan Rosen (1909–1995) and Boris Podolsky (1896–1966) imagined an experiment now called the EPR experiment, in which two particles are produced in superimposed state. For example, when a particle and its anti-particle combine, the two resulting photons must have exactly opposite polarization. The photons leave the site of the annihilation and travel out into space. If one observer detects the polarization of his photon to be 'up', then he knows instantly that the observer of the second photon will measure her photon to be 'down'. But we can make the separation between these observers so large that a light signal does not have time to travel between the observers to guarantee that the other photon is opposite. Since information cannot travel faster than light, we are left with the two photons in what is called an entangled state at macroscopic distances, but nevertheless defined as a single quantum state. Information apparently travels faster than light, which is not possible, or the description of the states of each photon carries more hidden information about measurement outcomes than allowed by quantum mechanics, which means that quantum mechanics is an incomplete theory. How can we possibly decide which is which?

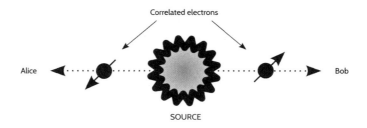

The Einstein-Podolsky-Rosen (EPR) Experiment places two particles in a single quantum state so that measurement of one particle tells you exactly what state the other particle is in, seemingly violating the relativity principle that information cannot travel faster than light.

In 1964, John Stewart Bell (1928–1990) proved that the second interpretation, that quantum mechanics was incomplete, was also incorrect. Bell's Theorem states that 'No physical theory of local Hidden Variables can ever reproduce all of the predictions of Quantum Mechanics.' The test of this theorem was ingeniously developed by Bell himself through a test of the predicted polarization of entangled photons assuming local realism is true or assuming that quantum mechanics' non-local realism is true. He developed a mathematical quantity called S, which was the total measured spin of the two particles in an experiment. The test involved an inequality, $|S| < 2.0$, which would be strictly obeyed if there were indeed hidden variables, but would be violated if quantum mechanics were strictly obeyed.

Beginning in the 1970s, there have been a series of experiments of increasing sophistication to test whether the Bell Inequalities are satisfied. In 2015, 'conclusive' tests were published by B. Hensen at the Delft University of Technology which claimed $|S| = 2.42 \pm 0.20$, which is technically called a 12-σ result because the value is 12 times larger than the statistical variation in the data, and by Marissa Giustina at the Institute for Quantum Optics and Quantum Information in Austria which claimed an 11-σ result. The bottom line is that any hidden variables theory to correct quantum mechanics for incompletes, and to introduce

John Stewart Bell developed a simple theorem that tests whether quantum theory can be repaired by invoking a larger theory that provides missing information.

de Broglie's pilot wave idea, is ruled out by these experiments to at least the 11-σ level. This kind of statistical confidence corresponds to the measured value occurring about one time in more than one trillion trillion measurements if it were simply a random measurement error. Local Realism is very strongly disallowed, and the peculiar nature of quantum mechanics appears to be firmly established.

The conclusion from these tests is that particles and observers can remain in complex, entangled states that prevent observers from experiencing a common reality, and that quantum systems *do not* have fixed properties until the moment of observation. This behaviour is not caused by quantum mechanics being an incomplete theory that needs to be patched up with a superior theory that can provide the missing information. The choice of the question posed by the experiment at the instant of interaction with the detector (are you a wave or a particle?) determines the property of the particle, independent of the paths taken by the particle.

Key Points

- Quantum superposition states that a particle exists in an infinite number of possible states until the moment of observation, at which point only one state comes into existence. This is called the Copenhagen Interpretation of quantum mechanics.

- Until the moment of observation, a particle can be in a superposition of more than two states at once, which is called quantum superposition.

- Two particles can be in a single superimposed state, in which the measurement of the state of one particle immediately places the other particle in a definite state. This is the basis for quantum entanglement.

- According to Bell's Inequalities, Local Realism is violated by quantum mechanics because a photon or electron does not know whether to be a particle or a wave until the instant of measurement with the appropriate instrument, which irrevocably selects which state to manifest.

- Local Realism cannot be built into quantum mechanics using hidden variables to eliminate the need for information transfer to be faster than light, as is seemingly required in the delayed choice experiment.

CHAPTER 6

Applications of Quantum Mechanics

We have now seen how the basic formulation of quantum mechanics helps us describe the properties of atomic matter, but what are the practical consequences of this rather arcane knowledge? In this chapter we will look at only a handful of situations in which the principles of quantum mechanics make a huge impact on our technology and beyond.

QUANTUM TUNNELLING

One of the most interesting and far-reaching aspects of wave functions is that they do not come to an abrupt end in space but, depending on the system in which they are in, can diminish to zero over the span of the entire universe. This means there can be a finite probability for a quantum particle to be found in places where classical physics says they shouldn't be.

Diagram of a wave function (curve) that passes through a barrier spanning a region of space (grey bar) with two different thicknesses.

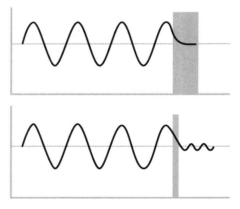

The figure shows a wave function for a quantum particle within a region of space that is bounded on the right by a barrier in energy that is higher than the energy level (horizontal line) of the quantum particle. The barrier also has a definite width in space. As it turns out from the mathematical solution for this particle, the wave function has a sinusoidal shape inside and outside the barrier region, but within the barrier the wave function decreases exponentially. The three solutions are smoothly joined to each other, which means that the particle's wave function on the right side of the barrier is non-zero unless the barrier energy is infinitely large. In ordinary Newtonian physics, the particle's wave function would be a constant value to the left of the barrier, but at the inside edge of the barrier the wave function would immediately fall to exactly zero because the particle has insufficient energy to surmount the barrier. The ability of a quantum particle to be found outside its classically limited region is called tunnelling. The time it takes a particle to escape from the permitted region is very long when the difference between the particle's energy and the barrier energy is large, and it is very short when this difference is small. One particular aspect of this tunnelling process is radioactive decay.

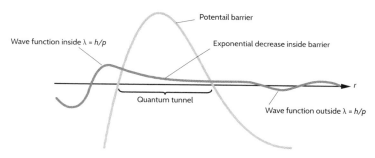

An example of tunnelling out of an atomic nucleus. Because wave functions do not come to an abrupt end, they can be found with non-zero amplitudes outside classically bound systems such as nuclei.

Particles found inside an atomic nucleus have insufficient energy to escape from the energy well caused by the strong nuclear force. However, these particles are defined by their own wave functions, and these wave functions extend throughout space, even outside the nucleus. Inside the nucleus, the wave function has a large amplitude because the probability to find the particles inside the nucleus is highest. However, the wave function can penetrate the potential energy barrier separating the nucleus from the rest of the atom due to the short-range of the strong nuclear force. Across this energy, the normally sinusoidal wave function takes on an exponential decay until its distance from the nucleus is large enough for it to be outside the domain of the strong force. From that point on, it resumes a sinusoidal shape but at a reduced amplitude. For most atoms, the time it takes to tunnel out of the nucleus is longer than the age of the universe and we perceive these atoms as stable. For other atoms, this tunnelling time can be measured in seconds, minutes and even billions of years. We call these atoms radioactive.

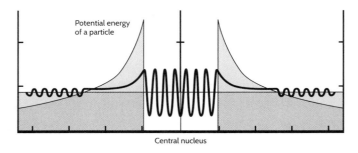

Quantum tunnelling of an alpha particle from inside an atomic nucleus.

Heavy nuclei above an atomic mass of about 200 can be considered as collections of alpha particles (the most tightly bound nuclei with 2 protons and 2 neutrons), which are found at various energy levels inside the nucleus similar to electrons within an atom. The most excited alpha particles, which are located close to the top of the nuclear energy well, can leak out of the nuclear energy well by tunnelling through a very small energy difference, which would still be forbidden in classical Newtonian physics. Depending on the energy difference between the alpha particle and the top of the energy well, the tunnelling will take a long time when this difference is large, and the nucleus will be essentially stable and non-radioactive; if the difference is small, the tunnelling will take a short time and the nucleus can be seen as radioactive. That is why the half-lives of some radioactive elements vary enormously, even though they may all primarily emit alpha particles during the decay process. For example, the half-life of bismuth-209 via alpha particle decay is 1.9×10^{19} years; for californium-240, it is one minute. Taking the long view, however, all atoms are ultimately radioactive and eventually subject to nuclear tunnelling, only over timescales longer than the current age of the universe.

A device called a tunnel diode relies on the quantum tunnelling of electrons at the junction between two materials to provide a weak current in a semiconductor.

Not only does quantum tunnelling explain radioactivity, but its practical engineering aspects have been known since the invention of the transistor in the late 1950s. An electrical device called the tunnel diode was invented by Leo Esaki in 1957 at the forerunner of Sony Corporation. Across a 10 nanometre-wide junction that should behave as a resistor when a voltage is applied one way, a weak current of electrons flows by the tunnelling process when a voltage is applied in the opposite direction. This device was the forerunner to the transistor developed a few years later by William Shockley.

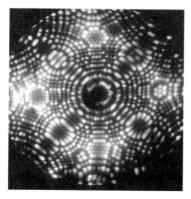

FIM view of platinum atoms made in 2006.

IMAGING ATOMS

Is it possible to photograph an atom? In 1951, Erwin Mueller (1911–1977) invented a new type of microscope that was, in fact, capable of imaging atoms by an indirect means. The Field Ion Microscope (FIM) used a tapered needle of tungsten at high voltage and detected the ions emitted by the tip to form an image of the individual tungsten atoms themselves.

FIM imaging was replaced in 1981 by the invention of the scanning tunnelling microscope (STM) developed by Gerd Binnig and Henrich Rohrer at IBM Zurich, for which they received the Nobel Prize in 1986. Using the tunnelling current between the tip of a needle and the surface of a material, the device moves horizontally across a surface, and the height of the needle is carefully varied to keep the tunnelling current constant. The mathematical surface formed from the height information is directly related to the presence and shape of the molecules and atoms beneath the tip's location. In fact, the atoms on the surface can be moved by the STM to create atom-sized machines.

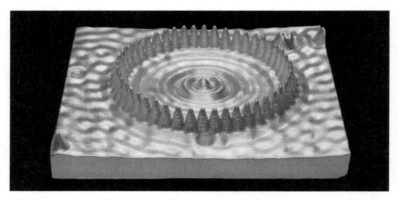

This 'quantum corral' was created by moving individual atoms using a STM instrument. The object in the centre shows where the wave functions of the individual atoms interfere to create a virtual atom.

IMAGING ELECTRON ORBITALS

Wave functions are mathematical objects that are related to the probability of finding an electron somewhere within an atom. They can be plotted to reveal the shape of these permitted regions, but the prevailing belief was that they could not be directly observed because they are merely clouds of probability and not actual, physical things like electric fields. In 2013, a remarkable image of individual electron orbitals was achieved indirectly by Aneta Stodolna of the FOM Institute for Atomic and Molecular Physics in the Netherlands. An instrument called a quantum microscope used laser beams to ionize atoms and trace their electrons from the point of interaction with the ionizing photons back to the detector. By mapping thousands of these interactions and tracks, the probability cloud for the hydrogen atom could be built up. These cloud maps, so far, match the quantum calculations, using Schrödinger's equation as exactly as the noise in the data will allow.

QUANTUM COMPUTING

Computers are based upon two-state, on-off switches, which are matched to the binary mathematical language used in all calculations. Computer software is simply a procedure for changing one set of these switches in the input state (question) to another set of switches in the desired output state (answer). The calculations require that the states, called bits, can be either fully on (binary value '1') or fully off (binary value '0'). Some programs, called algorithms, can partition portions of the calculation and run them concurrently with other calculations; a process called parallel computing, which greatly improves the speed of arriving at a final state. The algorithm can be thought of as taking a single

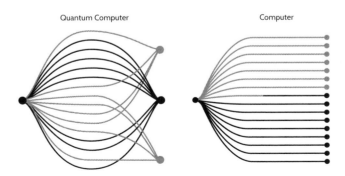

Example of a quantum computer connection between an input (problem) and output (solution).

pathway through all of the possible intermediate states between the input and output to get to the final answer. In the early 1980s, physicists Richard Feynman and Yuri Manin came up with a different approach to computing.

Normal computers using bits and algorithms were called classical computers because getting from an initial state (input) to a final state (output) required only operations in classical physics involving switches that are either off or on, and leading to a single, efficient algorithm for reaching a solution for a calculation. In a quantum computer, the algorithm takes all possible paths between the input and output state; and by the quantum process of superposition and entanglement, the most highly probable path becomes the most likely solution. To make this mode of computing possible requires not a 2-state bit but a 4-state quibit, where 1 and 0 are still represented, but superposed states between 1 and 0 also appear among the four possible states. Quantum computers are prone to noise, so they have to be super-cooled to cryogenic temperatures to prevent the quantum states among the quibits from collapsing into random noise. Nevertheless, by sampling all possible solutions between an initial and final state and weighing their probability, this method can be extremely fast for some classes of mathematical problems, such as those encountered in cryptography or in modelling the folding of proteins. The largest quantum computer to date is the IBM Q *System One*. It is the first integrated quantum computing system for commercial use and uses 20 quibits. Ironically, the creation of large quantum computers will have to confront the very difficult and sublime issue called the Problem of Time. There is currently no formal way to describe what it means for an observer to submit an input to such a computer and then later receive an output.

Key Points

- Quantum tunnelling accounts for the process of radioactivity, in which particles are emitted from nuclei despite not being permitted to do so according to classical Newtonian physics.

- The tunnelling time, or half-life, is determined by the difference in energy between the particle and the energy barrier.

- Individual atoms can be imaged both directly and indirectly using tunnelling techniques.

- Quantum computing allows input and output states to be placed into an entangled state that effectively samples all logical possibilities at once to find a solution.

- Advances in quantum computing hinge on answers to very deep questions in quantum physics, such as the origin and interpretation of time, as well as the distinction between the input and output states of a problem.

CHAPTER 7

Quantum Spin

In previous chapters we saw how the classical ideas of energy and momentum were carried over into a quantum theory of matter with little change. This is because, although each of these properties could be quantized in terms of small integer numbers such as 1, 2 or 3, they were nevertheless still meaningful ideas whether quantized within the atom or measured on the macroscopic scale of people, cars and planets. This is couched in the basic principle of quantum mechanics called the correspondence principle, which states that the behaviour of systems described by the theory of quantum mechanics reproduces classical physics in the limit of large quantum *numbers*. Although quantum physics is the province of atomic-scale systems, eventually quantum physics has to transition into ordinary Newtonian physics when the systems become large enough. This means, in principle, that every feature of the quantum world such as energy, momentum, mass, or charge, must be matched one for one with analogues in the macroscopic Newtonian world at large.

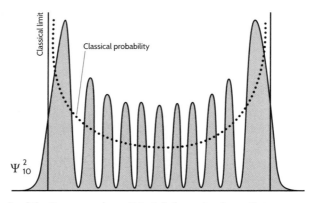

Example of the Correspondence Principle for a simple oscillator.

This figure shows a particle moving back and forth inside a box. The classical, Newtonian calculation represented by the dotted line shows the probability of finding it at some point between the two walls. As we might expect, this probability is highest near the walls where the particle bounces off the walls and spends most of its time. The quantum calculation shows nearly the same thing, though the wave function can have a non-zero 'tunnelling' probability outside the walls of the box. Nevertheless, the wave function matches very nearly the prediction from a classical, Newtonian treatment of the problem, as expected from the Correspondence Principle.

Another way to check the Correspondence Principle is to calculate what the electron probability cloud (defined as $P(x,y,z) = \varphi(x,y,z)^* \; \varphi(x,y,z)$) should look like for a simple hydrogen atom when the electron is furthest from the nucleus – as shown in the figure above. These so-called Rydberg atoms have their electrons in very high energy levels, E_n, for which the electron is at a distance of 1 micron – nearly 100 times the normal 'ground state' of such an electron in a hydrogen atom. The figure shows that the calculated wave function for a hydrogen atom in an $n = 15$ Rydberg state looks very nearly like a classical 'planetary' Bohr orbit for the electron.

QUANTUM SPIN

Even during the heyday of the quantum revolution during the mid-1920s, there seemed to be more going on with atomic physics than just the simple ideas of energy, momentum and angular 'orbital' momentum. By 1925, the detailed spectroscopic study of spectral lines within the framework of the ordinary quantum numbers for energy, momentum and angular momentum seemed to indicate that there was something more going on with electrons in these specific quantum states, defined by the three numbers n, l and m. Wolfgang Pauli (1900–1958) devised a mathematical scheme in which the quantum state of each electron in an atom was represented by four quantum numbers and not three. Two electrons could find themselves in the same quantum state represented by the first three numbers, n, l and m, but would have to have a different fourth quantum number, which led to the exclusion of any other electrons from the same four-number state. This became known as the Pauli Exclusion Principle. Pauli recognized that the first three quantum numbers were related to the energy and the orbital angular momentum of the electron, but could find no physical basis for explaining the fourth quantum number.

WOLFGANG PAULI

Wolfgang Pauli (1900–1958) worked under Arnold Sommerfeld and received his PhD in 1921 for his work on the quantum theory of the diatomic hydrogen atom. Three years later he proposed that there would be a fourth quantum number, spin, and this led to the principle of exclusion for electrons. He borrowed Werner Heisenberg's matrix mathematics to show how electron spin would be represented by a 2×2 matrix, which solved the problem of how to incorporate spin into Schrödinger's mathematics for quantum states. Pauli also discovered the neutrino in 1930. His lucid dreams were studied by Carl Jung, which led to a long series of famous letters between Jung and Pauli. In 1958 he died of pancreatic cancer.

Wolfgang Pauli (centre) with Paul Dirac and Rudolf Peierls.

A simple experiment performed in 1922 by German physicists Otto Stern (1888–1969) and Walther Gerlach (1889–1979) had all but revealed that it was the electron itself which was responsible for this fourth quantum number. They passed a beam of silver atoms through a magnetic field and discovered that the atoms favoured two orientations, defined by the orientation of their outermost electron. The prediction from the simple Bohr–Sommerfeld model would have shown a single orientation since no additional properties of the electron beyond its charge and mass were treated by the model. The fact that the electron manifested two states was interpreted to mean that there was an additional quantum number, s, that could have exactly two possible values: $s = +\frac{1}{2}$ and $s = -\frac{1}{2}$. The missing interpretation of what this meant was provided by the Dutch-American physicists George Uhlenbeck (1900–1988) and Samuel Goudsmit (1902–1978), who described this as the intrinsic spin of the electron, much as a planet rotates upon its axis while orbiting the sun. To see how the electron's spin quantum number changes the orbital arrangements of electrons in atoms, we have to take another look at the definitions of the quantum numbers n, l and m.

The three quantum numbers apply to each electron so that for a given energy level defined by the principle quantum number n, its values can be an integer like $0, 1, 2, 3....\infty$. The higher the value for n, the further is the electron orbital from the nucleus and also the closer the electron energy is to the ionization limit of the atom at $n = \infty$. In the terminology of chemistry, each value for n represents a different electron shell. Next, we have also seen how the total orbital angular momentum of the electron is given by the quantum number l. It describes the shape of the electron's wave function. The possible values for l can include 0, 1, 2 and up to the limit of $(n - 1)$, where n is the principle quantum number for the electron in its shell. Each value for l is called a subshell. Each substate defined by l also corresponds to a specific spectroscopic designation for a series of spectral lines represented by the letters s, p, d, f, etc.

George Uhlenbeck and Samuel Goudsmith described the intrinsic spin of electrons.

These letters were used by spectroscopists to describe what certain spectral lines looked like so that s = sharp, p = principal, d = diffuse and f = fundamental, for example. The l = 2 subshell is also called by chemists and spectroscopists the d-subshell.

The third quantum number, m, is called the magnetic quantum number and represents the projection of the orbital angular momentum along the z-axis of the atom. It can have values from - l to + l in integer steps. For example, for the p subshell with l = 1, the values for m can be only -1, 0 and +1. In general, the maximum number of subshells is given by 2l + 1, so that for the d-subshell with l = 2 we can have at most five subshells.

In the figure on page 97, the electrons find themselves in the '2p' energy state on the left and defined by the quantum numbers n = 2, l = 1. Historically, m is called the magnetic quantum number because, without an applied magnetic field, all of the electrons will be seen together in the same l-subshell as shown in the figure on the left – a condition called degeneracy. When a magnetic field is applied, the subshells sort themselves out into their separate m-levels as shown on the right, with at most two electrons in each subshell. This splitting of the l-shells leads to a separate set of spectral lines in a process called Zeeman splitting, first observed by the Dutch physicist Pieter Zeeman (1865–1943) in 1896. Two years later, the Irish physicist Thomas Preston (1860–1900) discovered that when some atoms were placed in very strong magnetic fields, even the Zeeman Effect seemed to lead to a further doubling of some spectral lines – a phenomenon later called the anomalous Zeeman Effect.

When a magnetic field is applied, the energy states on the left are split into subshells determined by the m quantum number, such that no more than two electrons can be found in each subshell. What Preston

Pieter Zeeman explored the change in atomic spectra when magnetic fields are applied, leading to the discovery of subshells in electron orbits.

discovered shortly before his death is that subshells can be further split into at most 2 additional quantum states. This splitting was finally explained by Pauli by assuming that the 2 stated derived from a fourth quantum number that obeyed the rule that 2 =

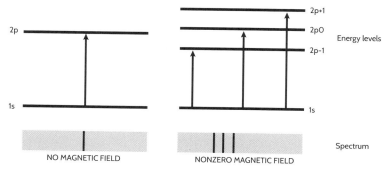

An example of the Zeeman Effect.

2s + 1 so that each electron could have at most 2 internal quantum states for which s = ½. The full collection of four quantum numbers n, l, m and s, now define the complete quantum state of an electron in an atom and also define a specific atomic orbital. Each orbital is a complete solution – $\varphi_{n,l,m,s}(x,t)$ – to Schrödinger's equation for the specific quantum state of an electron.

Although the new quantum number is called spin, and seems

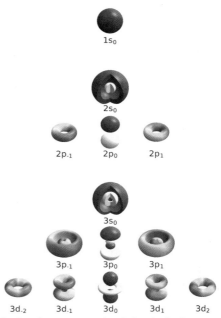

The first few atomic orbitals for the hydrogen atom for n = 1, 2, 3 showing the substates given in spectroscopic notation where $3d_2$ indicates the substate with n = 3, l = 2 and m = -2.

to suggest that electrons act like spinning planets, this phenomenon is in fact quite unlike ordinary rotation in space. The electron's spin results in what seems to be the apparent movement of a charged current, resulting in a weak magnetic field for the electron, called the magnetic moment. However, as a point particle, the electron has no physical extension that can be in rotation without exceeding the speed of light limit for physical objects. Also, there are only two allowed rotation states for the electron, determined by s = +½ and s = -½ such that in the former case the spin axis of the electron is in what might be termed the 'up' position and in the latter case in the 'down' position, but in each case the total magnitude of the spin |s| is exactly ½. Unlike the other quantum numbers for which the Correspondence Principle allows a classical interpretation for large enough quantum values, spin only comes in exactly two cases for particles such as electrons. This now opens the door for classifying elementary particles in terms of their intrinsic spin as well as their mass and charge.

BOSONS AND FERMIONS

The intrinsic spin of a collection of non-interacting particles leads to two types of statistical treatments of their properties as an ensemble. Normally, we have gases made from individual atoms and molecules that define density, temperature and pressure in the usual forms of 'gas laws' such as Boyle's Law (PV = constant, where P is the pressure of the gas and V the volume), or the Ideal Gas Law (PV = nRT, where T is the absolute temperature, n the number of moles of gas and R the ideal gas constant), which have been known since the 19th century. The discovery by 1925 that electrons have an intrinsic spin of ½, and an earlier discovery by Satyendra Bose in 1924 that photons also had an intrinsic spin of 1, led to two new statistical theories of gases consisting of pure spin-1 particles called bosons, and spin-½ particles called fermions.

The statistics of fermions together with the Pauli Exclusion Principle, which states that two fermions can be in the same quantum state only if they have opposite spins, leads to a number of important applications, especially in astrophysics. Under the extreme conditions of high temperature and pressure, ordinary atomic gases become plasma-rich in free electrons. These electron gases under further compression in volume reach a state in which the available quantum states for these fermions rapidly fill up. This generates what is called a degeneracy pressure that eventually resists further compression. The relationship between pressure and temperature also changes so that increasing the temperature of the electron gas no longer leads to an increase in the total pressure, which is now completely dominated by the quantum effects of the fermion gas under the Pauli Exclusion Principle. Systems in which degeneracy pressure are the dominant source of stability under the force of gravitational collapse include brown dwarf stars, white dwarf stars and neutron stars.

The statistics of bosons lead to the correct counting of photon states in a pure black body spectrum and an explanation for the Planck quantum method discovered by Bose. Other systems that incorporate Bose–Einstein statistics include Bose–Einstein condensates, which were discovered in 1995, though predicted by Einstein in 1924 and by Fritz London (1900–1954) in 1938. Einstein proposed that any particle that had a spin of 1, including helium atoms, would behave as a gas that would condense at low temperature to its lowest possible quantum state, forming a new phase of matter.

Satyendra Bose developed the statistical theory of integer-spin particles now called 'bosons'.

Key Points

- The Correspondence Principle states that for sufficiently large quantum numbers, a quantum system begins to resemble a classical system.

- Quantum spin is a fourth quantum number that represents the intrinsic spin of electrons, which may be either +½ or -½.

- Quantum spin leads to the Pauli Exclusion Principle so that at most two electrons can be in the same quantum state only if their spins are opposite so that s = +½ or s = -½.

- Fermions are elementary particles for which the magnitude of their spin is ½. Bosons are particles whose spins are integers such as 0, 1 or 2.

- Bose–Einstein statistics treat the states of spin-1 bosons while Fermi-Dirac statistics treat the states of spin-½ fermions.

- The Pauli Exclusion Principle produces degeneracy pressure when collections of fermions are compressed to high density, such as in neutron stars or white dwarfs.

CHAPTER 8
Relativistic Quantum Theory

During the 1920s, much of the excitement in physics centred on the refinement of quantum mechanics, but there was a fly in the theoretical ointment and it simply would not go away. No one really knew how to handle both light and matter in a completely quantum mechanical way. Even though matter and light were now both quantized with the help of Planck, Einstein and de Broglie, the actual mechanism by which quantum particles produced light remained obscure. In fact, the light

Gilbert Lewis was a chemist whose innovation of the term 'photon' is now universally used to describe the quantum of electromagnetic energy.

quantum wasn't even given a proper name until 1926, when the chemist Gilbert Lewis (1875–1946) proposed that it be called the photon.

The problem theoreticians faced was to craft a sound, logically seamless, mathematical theory for the following atomic phenomenon: 'The electron absorbs a photon and jumps to the next highest energy level, then after a brief period of time it falls back to a lower-energy state emitting a photon.' This is a deceptively simple process to express in words and to measure in a physics laboratory, but a very difficult one to describe mathematically. The first step was easy. The most difficult part had to do with the second step. Even in a perfect vacuum without a pre-existing electromagnetic field, an excited electron will jump to the lowest state available to it in an atom – a process called spontaneous emission.

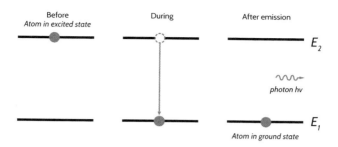

Before Atom in excited state	During	After emission

E_2

photon hv

E_1

Atom in ground state

Electrons do not remain in an excited state but jump to a ground state even though both are stable solutions of Schrödinger's Equation.

SPONTANEOUS EMISSION

Spontaneous emission was first described by Einstein in 1917 in a paper 'On the Quantum Theory of Radiation'. It was an idea that flatly contradicted the later predictions of quantum theory because the excited energy state of an electron is a perfectly acceptable solution to Schrödinger's equation; why didn't an electron just stay in its excited state forever? Einstein found that, upon closer scrutiny, the mathematics that described spontaneous emission required a peculiar assumption; it is caused by the forced emission of the electron due to a fictitious electromagnetic field. To make matters more complicated, this fictitious field would have to be present everywhere and at all times, even in otherwise perfectly empty space! Could Aristotle, Huygens and Maxwell have been right after all? Are we once again faced with the inescapable conclusion that space cannot be empty?

Paul Adrien Maurice Dirac (1902–1984) answered these questions by showing that spontaneous emission occurs because the number of photons in the electromagnetic field constantly changes due to Heisenberg's Uncertainty Principle, even when the field is absent! There is a built-in uncertainty about just how many photons are present at any frequency: a zero-point fluctuation in the lowest energy of that vibration frequency. Imagine trying to weigh yourself every morning, but with someone sneaking into your bathroom and fiddling with the setting. One morning you weigh 70 kilos, the next morning you weigh 75, and after that 65, 68, 80 and so on. Even if you didn't stand on your scale at all (equivalent to the electromagnetic field being shut off), the readings would show a fluctuation of +5, -5, -3, +10 kilos, and so on. How would you adjust your diet to make certain you didn't really gain or lose weight? This is the dilemma faced by the electron as it attempts to remain in its excited state. Inside the atom, the electron moves within the electromagnetic field of the proton. This field undergoes zero-point fluctuations that stimulate the electron to emit a photon of light and then cascade down into whatever lower energy level in the atom is unfilled.

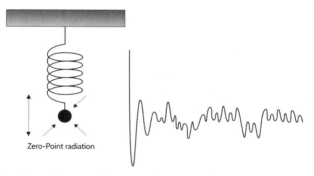

Zero-Point radiation

Zero-point fluctuations are changes in the energy of the physical vacuum state due to the appearance and disappearance of virtual particles, which cause changes in the physical behaviour of all quantum particles such as electrons.

PAUL DIRAC

Paul Adrien Maurice Dirac (1902–1984) graduated with a Bachelors degree in engineering in 1921 at the University of Bristol, but was unable to find a job during the post-war depression and returned to college. In 1923, he received a Bachelor of Arts in Mathematics. He studied general relativity and quantum physics, and was the first known doctoral student to complete a thesis on quantum physics in 1926. His work on relativistic quantum theory and quantum gravity formed the foundation of all future research in these areas, leading to his development of relativistic quantum field theory and the theory of 'Dirac Holes' in 1930. He shared the Nobel Prize in Physics in 1933 with Erwin Schrödinger 'for the discovery of new productive forms of atomic theory'. He was known as a man of few words, and a unit called the Dirac was humorously named after him; it is a speaking rate of one word per hour.

Paul Dirac.

What Einstein's investigation of spontaneous emission demonstrated was that even a region of space shorn of its classical electromagnetic field acts as if it still had an irreducible energy of $E = h\nu/2$, where ν is the frequency of the light wave, and h is Planck's constant. This energy is present throughout space and can yield some amazing effects. One of these is the so-called Casimir Effect, discovered in 1948 by the Dutch physicist Hendrik Casimir (1909–2000). Also called the vacuum force, this phenomenon occurs whenever two conducting plates are placed face-to-face. Maxwell's equations dictate that only certain modes of the zero-point energy can be maintained between the plates. This leads to less total vacuum energy between the plates than outside the plates, and it is this difference in pressure on the plates that drives them together.

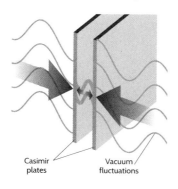

Casimir plates   Vacuum fluctuations

The Casimir Effect is an example of the zero-point energy of the quantum vacuum.

Spontaneous emission is not just some special process operating under unusual circumstances; it is the very source of the light we see from the sun. All of the atoms in the solar photosphere experience excitation, and de-excitation through the action of spontaneous emission.

The vacuum state can also be affected by changes in the properties of an enclosing, conductive cavity. Experiments by Randall Hulet in 1985 and others have demonstrated that an atom inside a small metallic cavity has a very different spontaneous emission rate than a similar atom outside the cavity in the unbounded vacuum. When this possibility was first discovered by Karl Drexhage in 1970, some theoreticians argued that this was impossible. The conventional wisdom was that spontaneous emission, like a human's fingerprint, was unalterable. A physics textbook *Fundamentals of Modern Physics* published in 1962 insisted: 'the transition rate for spontaneous emission is an inherent characteristic of the atom and is not influenced by the environment in which the atom is placed.' But this assertion was deleted in later editions.

RELATIVITY AND QUANTUM MECHANICS

When a new theory is being developed, a remarkable dynamism sometimes occurs between its developer and the mathematics being used. Although the basic ideas and thrust of the theory come from the mind of the theoretician, the theoretician can be led along seemingly logically consistent paths, to conclusions that defy all conventional thinking. Dirac explained his own credo: 'One should allow oneself to be led in the direction which mathematics suggests ... one must follow up a mathematical idea and see what its consequences are, even though one gets led to a domain which is completely foreign to what one started with.' We are, after all, human and sometimes it is difficult to believe the truth of what the mathematics is revealing. Dirac, himself, lost his nerve when faced with his next mathematical discovery.

In 1928, Dirac combined what was known about the properties of an electron – its fundamental charge, mass and spin – with Schrödinger's quantum theory and Einstein's special relativity to fashion a new relativistic theory of quantum mechanics. This was needed because the effective speeds of electrons in the atoms of heavy elements were significant fractions of the speed of light, and this also seemed an inevitable extension of quantum theory – in much the same way that Newton's physics and Maxwell's electrodynamics had also been recast as relativistic theories. Such was the deep intuition and insight of Dirac that he almost literally wrote down the correct relativistic replacement for Schrödinger's equation in one draft.

$$\left(\beta mc^2 + c \left(\sum_{n=1}^{3} \alpha_n p_n \right) \right) \psi(x,\ t) = i\hbar\ \frac{\partial \psi(x,\ t)}{\partial t}$$

where the quantities α and β are the electron spin matrices and p is the electron momentum operator: one for each space direction.

ANTIMATTER

Theories that are consistent with both relativity and quantum mechanics must invariably make the same prediction for the energy of a particle. But what is strange about this is that Dirac's relativistic theory predicted the existence of both positive and negative energy states for matter. This must happen because in relativity, it is not the total energy, E, of a particle that is conserved, but the square of its energy given by $E^2 = p^2c^2 + m^2c^4$ where p is the momentum of the particle, m is its mass and c is the speed of light. This means that two solutions exist for the energy of a particle with a mass, m, and a momentum, p: $E(+) = + \sqrt{p^2c^2 + m^2c^4}$ and $E(-) = -\sqrt{p^2c^2 + m^2c^4}$ corresponding to the positive and negative square roots of E. There is nothing especially magical about negative energy states. Earth moves in the gravitational field of the sun and, for this reason, is in a negative energy state – i.e. it's a bound 'particle'. The electron moves within the electromagnetic field of the atomic nucleus, so isn't the electron's negative energy state $E(-)$ just a reflection of it, too, being a bound particle like a planet? Not in this instance. If you removed the electron from the atom, the nuclear influence which defined the p^2c^2 term certainly vanishes, but you are still left with the energy solutions $E(+) = + mc^2$ and $E(-) = -mc^2$ which represent the irreducible rest masses, m, of the electrons – no matter where the electron is located in space.

Since the negative energy states still have a lower energy than the positive ones (because a negative quantity is always smaller than a positive one), and since electrons will seek the lowest energy state available to them, what prevents all electrons in the $E(+)$ state from instantly rushing to fill all of the negative energy $E(-)$ states? Why haven't all forms of matter in the universe vanished into the mysterious world of negative energy states? The physical world has indeed managed

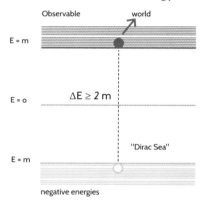

The Dirac vacuum is filled up by anti-electrons. However, a photon with a suitable energy (2mc²) can create an electron-positron pair.

to avoid this fate, so to make the new generation of relativistic quantum theories consistent with the real world, Dirac proposed in 1930 that the negative energy states of the vacuum were already occupied.

The innocuous energy equation had in one fell swoop led to a revolutionary new conception of empty space. The quantum mechanical vacuum was not empty but filled to the brim with an infinite number of anti-particles already occupying all of the negative energy states. Occasionally, some of these negative energy states would be vacant and the resulting Dirac Hole would be seen as a positively charged particle. It is at this point that Dirac departed from his own credo in interpreting mathematics. Even though the masses of these Dirac Holes would always turn out to be that of the electron, Dirac thought that these particles were actually the positively charged protons, which were 1864 times heavier than the electron. The following year, Hermann Weyl was able to prove mathematically that Dirac Holes had to have exactly the same mass as electrons. This meant that a particle identical to the electron, except for its opposite charge, should exist in Nature if Dirac's relativistic theory was the correct one.

In 1932, Carl Anderson (1905–1991) discovered the anti-electron, one year before Dirac's paper appeared in print, but the significance of this discovery was not immediately understood until after the Dirac Hole had been predicted. In 1933, Patrick Blackett (1897–1994) and Giuseppe Occhialini (1907–1993) succeeded in showing that the positron was, in fact, the Dirac Hole and interpreted the electron falling into an unoccupied negative-energy hole as the collision between an electron and a positron. The collision resulted in their mutual annihilation and the production of two gamma rays, carrying away a total energy of $E = 2mc^2$.

The union of the two great foundations of physics – quantum mechanics and special relativity – had led to the introduction of anti-matter as a coequal inhabitant of the physical world. But the

Carl Anderson discovered the anti-electron from studies of cosmic ray tracks in photographic plates.

discovery of the positron did far more than merely add a new particle to join the electron, proton, neutron and photon. It provided a confirmation to Dirac's hypothesis of the contents of empty space. Empty space in the quantum view was filled with an infinite number of anti-particles in a so-called negative energy sea. These particles cannot be detected, and do not behave in any way like the older notions of the ether. But under certain circumstances, pieces of this sea can be made to enter the real world as a new particle; the positron or even more massive ones like the proton itself. The vacuum state turned out to be a very crowded arena indeed. It was beginning to seem as though Aristotle and Descartes may have had the right idea after all. Nature really did seem to abhor a vacuum – at least one defined by a mathematically perfect emptiness. Nature preferred to fill the vacuum with ghostly forms of failed matter.

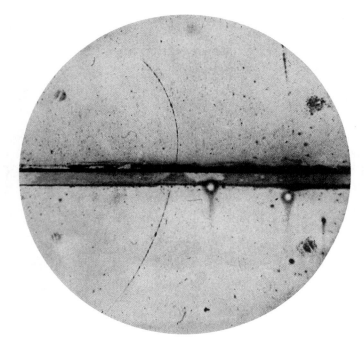

Cloud chamber photograph of a positron.

Key Points

- Spontaneous emission in a vacuum with no electromagnetic field suggested to Einstein and others that Heisenberg's Uncertainty Principle leads to a new zero-point energy for empty space.

- The Casimir Effect is a physical demonstration of the presence of a new zero-point electromagnetic field embedded in the vacuum.

- Special Relativity combined with quantum mechanics led to the prediction of anti-electrons, and their eventual discovery as the positron.

- Dirac's negative-energy sea had to be filled up so that normal matter would not decay into these new states and become unstable.

- An electron falling into a lower-energy Dirac Hole is physically interpreted as an electron and positron colliding and producing a pair of gamma-rays equal in energy to the combined rest masses of the electron and positron.

CHAPTER 9
Quantum Field Theory

In 1924, at about the same time that renewed attention was being directed towards developing a true quantum theory of light, another perspective on the subject of how atoms emit light made its way into the literature. Niels Bohr, Hendrik Kramers (1894–1952) and John Slater (1900–1976) made a bold proposal that completely rejected the quantum character of light. In a last bid for a purely classical description, they believed – in line with Maxwell and many others – that light was a strictly continuous phenomenon. Any of its apparent quantum characteristics involved the way that light interacted with matter. They conjectured that energy conservation doesn't strictly hold for individual atomic events, but is valid only in some average sense when many interactions are added. Something which they called a virtual radiation field was associated with each atom, which contained photons carrying every possible energy difference among the energy states in each atom. If an atom could produce a spectral line at 6519 Angstroms, there would be a virtual photon with exactly this required energy just waiting to be born under the right conditions.

VIRTUAL PHOTONS

Although this model was eventually disproved by experiments in 1925 carried out by Walther Bothe (1891–1957) and Johannes Geiger (1882–1945), the idea that there might exist a virtual radiation field violating strict energy conservation would soon find its way back into the mainstream of physics. Together with Heisenberg's Uncertainty Principle, it would, in fact, become the

Walther Bothe experimentally disproved the Kramer-Slater theory of the virtual radiation field in atoms.

next most significant new element added to understanding space and the physical vacuum. As it turned out, the virtual photons were far more numerous and varied in energy than is implied by the few thousand spectral lines in a typical atom.

In the newly developed theory of the quantum field, the electrostatic Coulomb field that causes like-charged particles to repel and dissimilar-charged particles to attract, can be thought of as a swarming hive of virtual photons continuously being emitted and absorbed by the electron. If we were to calculate the total energy of the photons and electrons at each instant in time, we would find that energy conservation is badly violated during that brief moment when the electron has emitted the photon and is preparing to reabsorb it:

Before: $E = mc^2$
During: $E = mc^2 + h\upsilon$
After: $E = mc^2$

Provided that the photon doesn't exist longer than the time specified by Heisenberg's Uncertainty Principle, Nature won't allow us to observe this violation of energy conservation.

To understand how this works, consider the following: if the energy of a photon is given by $E = h\upsilon$ and if the Heisenberg Uncertainty Principle says that $\Delta E \times \Delta t < h/4\pi$ where h is Planck's Constant (4.1×10^{-15} eV s). An energy uncertainty of ΔE corresponds to a measurement duration of $h/(4\pi\Delta t)$. This means a photon with an energy of $\Delta E = 1$eV can last no longer than $\Delta t = 3 \times 10^{-16}$ seconds before it must be absorbed by the electron to escape detection. During its fleeting lifetime, travelling at the speed of light, it can move a distance of about 300 angstroms and be absorbed by some other electron. Since atoms are only a few angstroms in diameter, virtual photons can transmit the influence of the electron over a great distance. Upon absorption by another electron or charged body, the virtual photon vanishes and imparts a change-in-momentum kick that is felt as the Coulombic force.

The exchange of a virtual photon between two electrons causes each to experience a momentum 'kick' that is viewed as the electromagnetic force between them.

Another more light-hearted nickname for this process is the Quantum Embezzlement Principle. You can embezzle one penny for a year, but you cannot embezzle a million dollars for more than a second without getting caught. This economic analogy also points out that Nature, like the Tax Inspector, is basically fair. Nature and the Tax Inspector will allow you time to balance your accounting books by a specified time, and not demand that the books be balanced to the penny at each step in the accounting process. If this view of Nature is correct, there is no limit to the energy these virtual photons may embezzle from Nature. Some of them can take minutes, years or centuries to pay back their debt, and be free to roam to unlimited distances. We now have an explanation for why the Coulomb force has such a large range! The least energetic virtual photon can travel feet, miles, and even light years before it has to be reabsorbed by an electron somewhere else to balance nature's energy accounting books. Trouble arises when we look at what happens for very short distances. Since there is also no limit to how short a wavelength virtual photons may have, some could, in principle, carry more energy than the sun produces in a billion years, which is the tidy sum of 10^{50} ergs, provided that they don't live longer than 10^{-77} seconds or travel farther than 10^{-67} cm from the electron that gave birth to them. Because the energy carried by these short wavelength virtual photons can even be infinite, there must be something that steps in to conceal this 'infinity' disaster from view with fantastic precision, otherwise the mass of the electron would be measured in, say, tons, not in increments of 10^{-28} grams!

THE PHYSICAL VACUUM

The first few years of the 1930s were marked by a number of imaginative solutions to the electron's 'infinity problem' (as it came to be known). Not long after Dirac discovered the positron and the properties of empty space, he uncovered yet another ingredient to the vacuum state. In addition to the virtual photons of the electromagnetic field, and the great sea of negative energy states and holes, Dirac had shown that the physical vacuum was also a rich cauldron of virtual positron–electron pairs that were continuously appearing and disappearing everywhere in space, in strict accordance with Heisenberg's Uncertainty Principle. The first theoreticians to add these pairs into the mathematical stew of calculating the electron's mass were Wendell Furry (1907–1984) and J. Robert Oppenheimer (1904–1967) in 1934, and a few months later Wolfgang Pauli and Victor Weisskopf (1908–2002). They discovered in their mathematical deliberations an interesting and unexpected result: the pairs that appeared and disappeared in an electromagnetic field could contribute a slightly negative energy to the field. What the pairs did was to produce an effect now called vacuum polarization in the space surrounding the electron.

J. Robert Oppenheimer conducted some of the earliest mathematical calculations using a primitive version of what later became the theory of quantum electrodynamics.

The positive charge on the positron in the pair is attracted by the electron, while the virtual electron's negative charge is repelled. From a distance, the positron cancels some of the electron's negative charge. The resulting strength of the electron's field would weaken at small enough distances. But this negative energy was not sufficient to balance the positive infinity of the electron's own electromagnetic field.

In the face of this by now all too familiar and discouraging result, Furry and Oppenheimer proposed, in desperation, that eventually a limit must be reached where the sum over all the energies for the nascent pairs can be naturally ended. For example, consider the simple harmonic series $1 + \frac{1}{2} + \frac{1}{3} + \frac{1}{4} +$ etc. It diverges to infinity, but if you only add the first one million terms you will get the finite answer 14.3927. Furry and Oppenheimer suggested that a suitable scale to end the pair calculation might be the Compton wavelength of the electron defined as $\lambda = h/mc$ where h is Planck's constant (6.6×10^{-34} joules sec), m is the electron mass (9.1×10^{-31} kg), and c is the speed of light (3×10^{8} m/s) and corresponds to a distance of 0.0024 nanometres.

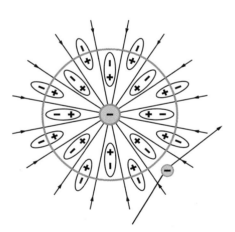

When an electron is placed in the vacuum, the virtual particle pairs align to polarize the vacuum and cause a slight change in the energy and charge of the electron.

Now the Compton radius of the electron was not supposed to be thought of as the size of some hard sphere. Instead, it would represent a boundary in spacetime beyond which conventional theories of relativity and quantum mechanics may break down, requiring some new theory to be used. By analogy, the speed of sound is not a physical barrier to airplane speed but is merely a marker in velocity where new physical phenomena such as shock waves make their appearance, which can destroy airplanes not specifically designed for surviving them. This approach was an entirely reasonable way out of the dilemma and many physicists tried various flavours of it to solve the infinity problem. As Walter Heitler (1904–1981) pointed out in 1961, there were no experiments that demanded this inner region of spacetime obey the same physical laws as the region outside what became known as the cut-off radius. Unfortunately, even when such cut-offs were included in the calculations of, for example, the electron's mass, the anticipated finite answers would nevertheless be wrong when compared to experiments. There was still some other ingredient that had been missing from the theories on which the calculations were based.

QUANTUM FIELD THEORY

By now, we have seen how the extension of relativistic quantum mechanics to include electron–positron pairs led to an enriched interpretation of what we mean by the term vacuum. The physical vacuum includes large numbers of virtual photons to explain the process of spontaneous emission. There were also electron–positron pairs with a combined mass of two electrons ($\Delta E = 2mc^2$) whose brief existence set by Heisenberg's Uncertainty Principle of about $\Delta t = (h/4\pi)(1/2mc^2) = 3 \times 10^{-22}$ seconds can have a significant effect on whatever particles or fields are present. However, none of these ingredients led to a theory capable of making a single valid prediction. By 1940, calculations could show at best that the mass of the electron would not become infinite at too rapid a rate as the term-by-term calculations progressed to their conclusion. This meant that perhaps all the talk about virtual photons and positron–electron pairs in empty space was nothing more than sheer fantasy.

There was a general malaise among theoreticians over the directions that Quantum Electrodynamics – or QED as it was now called – was taking them, and whether they could ever find their way back to calculations that had meaning in the real world. The infinity problem as it came to be known, caused many physicists to believe that quantum field theory was not even a very good theory for describing how electrons interact with fields. Weisskopf recollected that 'The filling of the vacuum and Dirac's theory of the positron – nobody really believed it at the time. And the theory was so ugly! ... and a lot of people just said: "To hell with it, I'm going to do some nuclear physics [instead]." Yet this new arena of research was a seductive field of study for theoreticians, who continued to dabble. Like prospectors searching for the mother lode, many felt very strongly that success was just a matter of making one more adjustment to the theory and finding one more clever mathematical technique for removing infinity.

Much of the research on QED came to a halt during the Second World War, when many physicists who had been so active in its early development were asked to turn their talents to developing the atomic bomb. Perhaps all the theoreticians really needed was some time off. The work on the atomic bomb certainly provided abundant time away from quantum field theory. When World War II ended and physicists returned to the bucolic settings of academe, the research began anew, and with greatly increased motivation. There was now a new piece of data, which could not be successfully explained by ordinary quantum mechanics. Only the dubious, infant theory of QED had been able to anticipate it.

THE LAMB SHIFT

At Columbia University in 1947, Willis Lamb (1913–2008) and his student Robert Curtis Retherford (1912–1981) announced the detection of a new kind of energy shift for the electron. They had successfully used certain microwave technology developed during the Second World War to measure the energies of two of the electron states in the hydrogen atom. Rather than finding them equal as predicted from calculations using Dirac's relativistic formalism, they found that actual measurements differed from the predictions by 4 parts in 10 million! This slight shift in the electron energy levels, beyond what ordinary Dirac theory required, was precisely the effect that Oppenheimer had been trying to calculate some 17 years earlier due to the interaction of the electron with its own electromagnetic field. Hans Bethe calculated the expected Lamb Shift, and his back of the envelope result agreed with the Lamb–Retherford measurement. Yet as

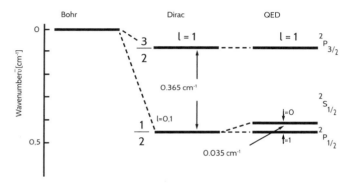

The Lamb Shift is a slight change in the energy of an electron inside an atom due to the effect of the virtual particles in the vacuum state.

Freeman Dyson reflects on this period of time, Bethe's calculation '…was a pastiche of old ideas held together by physical intuition. It had no firm mathematical basis. And it was not even consistent with Einstein's principle of relativity.'

FEYNMAN DIAGRAMS

By the end of the 1940s, thanks to the efforts of Julian Schwinger (1918–1994), Freeman Dyson, Sin-Itiro Tomonaga (1906–1979) and Richard Feynman (1918–1988), QED became the theory of choice for completely describing how charged particles like electrons and protons interacted among themselves, and with light. Feynman even developed a diagrammatic method for showing how a wide range of phenomena work. In a Feynman diagram, electrons are drawn as solid lines with arrows pointing forward in time, representing the history (called a worldline in relativity) of the electron as it moves through space. To show that photons are a different kind of particle involved in the transmission of the electromagnetic force, they are drawn as wavy lines. The intersection points are called vertices, and symbolize the instant at which the photons (electromagnetic field) interact with the particle. The ordering of events in time within the process becomes completely unimportant because in the spirit of these four-dimensional snapshots, the diagrams represent the entire process from start to finish in its totality.

For example, in the simple process of an electron interacting with a positron, the events ordered in time are as follows: The original electron is annihilated at point A by the positron, resulting in the production of two gamma rays. According to Feynman's reinterpretation of this process, an electron is scattered by the vacuum at point A and emits two photons. The electron then moves backwards in time to point C. The notion

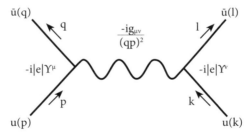

A Feynman diagram displays the mathematical terms that must be multiplied together to calculate the probability for a quantum process involving the exchange of virtual particles such as in this case the photon represented by the wavy line. They are not 'photographs' of what is actually occuring during the interaction.

that particles may move backwards in time, though bizarre, has a firm basis in mathematics. It had been known since 1942 that a positron moving forwards in time can be treated mathematically as an electron travelling backwards in time. Instead of two separate particles, the electron and the positron, there is only the one electron, which zigzags back and forth in time. By this way of thinking, the sudden appearance and disappearance of a positron–electron pair in space represents a single electron cycling endlessly between two instants in time in a closed loop.

As provocative as they initially seemed, Feynman diagrams were nothing more than symbols for various terms in an equation, and could just as easily be thought of in less provocative ways. In electronics we use specific, but arbitrary, symbols for representing resistors, capacitors and transistors. The symbols, like notes on a musical staff, only vaguely have anything to do with what the electrons are actually doing as they run around a circuit in a radio or a computer. Yet these symbolic patterns help us logically design devices of staggering complexity, involving millions of discrete components. In QED, they help us keep track of the terms we have to multiply together to get at our final result. Each element of the diagram is a particular factor that is multiplied with all the other factors and evaluated to determine how much the particular diagram contributes to the overall process. For example, the figure on the previous page shows two electrons interacting through their electrostatic Coulomb force, and is used following a specific set of rules developed by Feynman to construct the following formula, which has to be summed over all space to compute the 'amplitude' (i.e. probability) for this particular graph:

$$A = (-ie)^2 \bar{u}(q)\gamma^\mu u(p) \frac{-ig_{\mu\nu}}{(q-p)^2} \bar{u}(l)\gamma^\nu u(k)\delta(p+k-q-l)$$

Although the diagrams seemed not to follow the convention of quantum mechanics – that electrons do not follow neat, narrow trajectories through space – nevertheless, when Feynman and Schwinger compared their methods for calculating the outcome of the same process, the results were identical. Each element of the Feynman diagram could be matched one-for-one with a term or a factor in the complex mathematical formalism used by Schwinger. Whereas it would take weeks to derive an answer using Schwinger's formal mathematics, for Feynman and his diagrammatic methods the same computation took only a few hours.

RICHARD FEYNMAN

Richard Feynman (1918–1988) grew up in Far Rockaway, New York, attended MIT as an undergraduate to major in engineering, and later Princeton University, where he received his PhD in physics in 1942. After his work on the Manhattan Project, he made major contributions to quantum electro-dynamics by applying

Richard Feynman.

path-integral techniques which led him to the development of Feynman Diagrams and simple diagrammatic rules for a variety of calculations in QED. At the California Institute of Technology, he worked on the parton model for the protons (a complement to the quark model) and issues in superconductivity. Along with Julian Schwinger and Sin-Itiro Tomonaga, he was awarded the 1965 Nobel Prize in Physics for his work on the foundations of QED. Later, Feynman uncovered the reason for the Challenger Shuttle failure in 1986 as a member of the investigatory commission. He died of liposarcoma in 1988. His last words were: 'I'd hate to die twice. It's so boring.'

On the experimental side, Dirac first used QED to calculate the magnitude of the anomalous magnetic moment (g-2) of the electron, which has served as a high-water mark for a succession of battles between theoreticians and experimenters. The calculation of this effect is tedious though mathematically well defined. In 2015, Toichiro Kinoshita and his group at Cornell University calculated the QED correction to the 10th order involving 10 photons, and obtained (g-2) = 0.001159652181643 ± 0.000000000000023. This involved evaluating over 12,000 Feynman diagrams. The experimental value determined by David Hanneke and his colleagues at Harvard University was (g-2) = 0.00115965218073 ± 0.00000000000028. The difference between the two is just one part per trillion.

Despite the precision with which QED allows calculations to be carried out, some theoreticians feel that QED is missing some vital ingredient; some key principle. There has even been some discussion as to whether QED is a theory at all as opposed to simply a collection of calculational techniques and patch-ups for them. Theoretically, the key ingredient to the success of this theory is the early introduction of a mathematical technique called renormalization, which eliminated the various infinities that plagued earlier versions of this theory.

Renormalization is a mathematical technique rooted in the idea that the mass and charge of the electron appearing in equations as m and q, called the bare mass and bare charge, are not the same as the mass and charge that are actually observed. The mass and charge you measure is modified by their interactions with the physical vacuum and the virtual particles that come and go within it. The bare charge is partially shielded by the appearance of the virtual electron–positron pairs. If the bare charge and the vacuum shielding were treated separately, they would lead to calculations that showed them to be infinite. However, because these two infinities are of opposite sign, they can be combined to give a finite result if you simply redefine the electron's charge as q(obs) = q(bare) + q(vacuum). A similar issue arises from the mass of the electron, so that after it is renormalized you get m(obs) = m(bare) + m(vacuum). By renormalizing the original quantum field theory in this way, you get the modern theory of QED. This technique led some physicists to consider QED an imperfect and deeply flawed theory, no matter how well its calculations seemed to match up with experimental reality. Renormalization looked like a patch-up and an ad hoc repair for making a flawed theory work successfully. Even Dirac, who had a passionate respect for self-consistent mathematical formalisms, noted in 1984 that, 'Many people are happy [with the success of QED] because it has a limited amount of success. But this is not good enough. Physics must be based on strict mathematics. One can conclude that the fundamental ideas of the existing theory are wrong. A new mathematical basis is needed.' In his book *QED: The Strange Theory of Light and Matter*, Feynman went so far as to call renormalization a 'dippy process' – nothing more than hocus-pocus, which he claimed has prevented us from finding a better theory of QED that is truly self-consistent, and without any infinities to begin with.

Nevertheless, it seems that Heinrich Hertz's 1894 vision has triumphed. Some 50 years before the formulation of QED, Hertz published *The Principles of Mechanics Presented in a New Form* in which he noted 'If we wish to obtain an image of the universe which shall be well-rounded, complete and conformable to law, we have to presuppose behind the things we see, other invisible things. [We have] to imagine confederates concealed beyond the limits of our senses.' But now we know that, under certain controllable conditions, we can bring these invisible confederates into the light, revealing them to be far more than mathematical hyperbolae. By colliding heavy nuclei together, the vacuum can be sparked to emit electron–positron pairs. Physicists Jack Greenberg (1927–2005) at Yale and Walter Greiner (1935–2016) at the Goethe University in Germany described how at an atomic number exceeding $Z = 173$, the electromagnetic field of such a nucleus is so strong that it is unstable to the spontaneous creation of virtual positrons. By colliding uranium atoms, such a state can be created for a moment, and the glimmering of the sparking QED vacuum can be detected.

So far, we have succeeded in completely eliminating the ether from electrodynamics. It has been replaced by the electromagnetic quantum field of virtual photons. We have also replaced the concern for action at a distance with, first, a local field, and then with interacting virtual particles. Enigmatic classical fields have become equally enigmatic quantum fields. Instead of mulling over the reason for action at a distance, we now get to puzzle over what virtual particles are in themselves. If this new reality is giving you something of a headache, you are in good company. Richard Feynman in his 1964 Messenger Lecture at MIT once said about quantum mechanics: 'I think it is safe to say that no one understands quantum mechanics. Do not keep saying to yourself, if you can possibly avoid it, "but how can it be like that?" because you will go down the drain into a blind alley from which nobody has yet escaped. Nobody knows how it can be like that.'

We have now reached what appears to be the final successor to Schrödinger's quantum theory, and to the improvements on it made by Pauli and Dirac. Following many tedious years of effort on the part of a dozen theoreticians, renormalization was destined to play a major role in the success of QED. However, one of the casualties in QED was the electron itself.

For QED to work, the electron had to be a mathematical point in space or it would violate the relativity principle, which would forbid us from defining an absolute frame of reference by merely looking at the shape of the electron. All hopes of using quantum electrodynamics to derive the electron's actual 'bare' mass or charge had to be given up thanks to the renormalization technique used to banish infinity from the calculations. Like the Five Postulates of Euclid, or the constancy of the speed of light for relativity, the value for the electron's charge and mass could not be predicted but had to be measured and then put into the equations by hand. The physicists had struck a Devil's bargain with Nature: if you don't want infinity as an answer, don't ask stupid questions like 'What does an electron look like without its electromagnetic field?' Since you can never 'turn off' the electron's field, it didn't seem like this arrangement was really that objectionable. What had happened, in fact, is that Nature had taught us that the masses and charges we had been using so indiscriminately in our quantum theories were not the things we actually measure but something far more abstract.

QED also absolutely required that the physical vacuum be filled with a bewildering sea of activity, forever hidden from direct view by Nature's Heisenberg bookkeeping. Without this new concept for the vacuum state, none of the mathematics in QED work, and a slew of experiments would continue to go unanswered by a single theoretical prediction. As it now stands, QED is one of the triumphs of physics. Its only other historical competitor is Newton's physics and its successor, special relativity. A delightful feature of QED, and in particular Feynman's diagrammatic methods for describing electrodynamic processes, is that we now have a mathematically accurate way of drawing a picture of the contents of what we think of as empty space and what we now have to call the physical vacuum!

As we will see in the next chapter, the QED theory for describing electrodynamics at the quantum level was greatly expanded between the 1950s and 1970s to include two additional forces in Nature which also played a role in subatomic physics: the strong and weak nuclear forces.

Key Points

- Quantum field theory is the relativistic theory for electrons and photons which describes fields of force as clouds of virtual particles that are exchanged subject to Heisenberg's Uncertainty Principle.

- Virtual photons can violate conservation of energy as they are being exchanged by electrons, but only for a period of time limited by the Heisenberg Uncertainty Principle.

- The Lamb Shift represents an interaction between electrons in an atom and the virtual particles and processes occurring in the physical vacuum.

- Feynman diagrams represent the possible virtual processes available to a system of electrons and photons as they interact with the physical vacuum. They are not, however, literal 'pictures' of how those interactions appear in space and time.

- Renormalization is a technique for removing infinities by redefining the charge and mass of an electron so that infinities cancel out, and this has led to the development of Quantum Electrodynamics (QED).

CHAPTER 10
The Standard Model

Following a non-stop series of discoveries during the 1950s and '60s, the essential ingredients to Nature were quickly uncovered at high-energy accelerator laboratories around the world. In the discussions to follow, I will refer to the masses of particles based on their energy equivalents in units of electron-Volts (eV) following Einstein's equation $E=mc^2$. However instead of expressing the mass as eV/c^2, which would technically be the correct thing to do, I will follow the convention used by physicists and only mention their equivalent eV. The division by c^2 to get the correct mass in kilograms is understood. The suffix MeV means million eV and GeV means billion eV.

The familiar electron (0.5 MeV) gained a partner called the neutrino, but then two more massive particles were discovered called the muon (105 MeV) and the tauon (1.777 GeV) with their own partner neutrinos. There were now three generations of these so-called leptons, each more massive than the next, but since 1975 no new generation of still-more-massive leptons have been discovered that are heavier than the tau lepton. Meanwhile, the proton and neutron were joined by dozens of other massive particles as accelerator 'atom smashers' created them in a steady stream of discovery extending since the 1950s to the present time. By the early 1960s, the quark structure of sub-nuclear particles had been postulated by Murray Gell-Mann (1929–2019) and George Zweig (1937–) in an effort to simplify the classification scheme for the burgeoning number of such particles discovered every year. There were two quarks called Up (U) and Down (D) that combined three at a time to form the proton (U+U+D), neutron (D+D+U) and other low-mass particles, followed by the Strange (S) and Charmed (C) quarks forming a second generation, and the still-more-massive Top (T) and Bottom (B) quarks rounding out the third generation in the quark family. The leptons and quarks were all fermi particles (fermions) carrying a spin of ½ and together with their anti-particles formed the matter content of what is now called the Standard Model. Despite considerable effort and expense, no new quarks have been discovered since 1995 that are more massive than the Top quark (173 GeV). Alongside the reduction of matter into a small collection of elementary particles came the rapid evolution of the mathematical description for the forces that control them. We have already seen how Maxwell

	QUARKS		LEPTONS	
Generation 3	t Top	b Bottom	τ Tau	ν_τ Tau-neutrino
Generation 2	c Charm	s Strange	μ Muon	ν_μ Muon-neutrino
Generation 1	u Up	d Down	e Electron	ν_e Electron-neutrino

In the Standard Model, six quarks and six leptons are the fundamental matter particles, which come in three generations, each more massive than the previous one.

created a unified picture of electric and magnetic fields. Once the basic rules of special relativity and quantum mechanics were established, the electromagnetic field was quantized, and by 1949 QED emerged. By 1955, two other forces had been identified which controlled various subatomic processes such as the radioactive decay of the neutron (the weak nuclear force) and the binding of neutrons and protons into the tight confines of an atomic nucleus (the strong nuclear force). But aside from some simple models about how these forces acted, nothing even remotely resembling a quantum theory with the scope of QED was yet available for either of these nuclear forces. There were some tantalizing clues, though, as to what such theories ought to look like.

YUKAWA THEORY FOR FORCES

Some 20 years earlier, Hideki Yukawa (1907–1981) proposed that the strong force could be produced by the exchange between the proton and neutron of a particle whose range would be dictated by Heisenberg's Uncertainty Principle. This Yukawa particle (actually three different particles called the π mesons or pions) was

Hideki Yukawa developed the first theory of the exchange of particles as the causes for the fundamental forces of nature.

eventually discovered by César Lattes (1924–2005) and Giuseppe Occhialini (1907–1993) in 1947. Yukawa went on to win the Nobel Prize for his work in 1949. A similar exchange process was considered as a model for the weak force, and by the 1950s the theoretical literature was rife with candidate field theories for both the strong and the weak forces, involving the exchange of massive particles that mediated the forces.

The idea was borrowed from the basis for QED in which a force-carrying, and massless, virtual photon was exchanged under the umbrella of Heisenberg's Uncertainty Principle, but now the particle was allowed to possess its own intrinsic mass. As a result, the mathematics provided by Yukawa showed that the range of the resulting force would take an exponential form:

$$F = f(x)e^{-\frac{x}{r}}$$

where r was the range of the force defined by Heisenberg's Uncertainty Principle such that $\Delta E = (h/4\pi)/\Delta t$, and where $\Delta E = mc^2$. The uncertainty in time, Δt, could then be related to the range of the force via the speed of light: $r = c\Delta t$, to get the relationship between the range and the mass of the exchanged particle:

$$r = \left(\frac{h}{4\pi}\right)\frac{1}{mc}$$

According to the Yukawa model, for a pion mass of 135 MeV or 2.4×10^{-28} kg, and with Planck's constant h = 6.6×10^{-34} kg m/s^2, and c = 3×10^8 m/s, the pion range for the strong force should be about 7.3×10^{-16} metres. This is comparable to the radius of an average atomic nucleus where the strong force would dominate to keep the neutrons and protons bound as a system against the tremendous electrostatic repulsion of the protons.

YANG-MILLS THEORY AND QUARKS

Then in 1954, Yang Chen-Ning (1922–) and Robert Mills (1927–1999) at the Brookhaven National Laboratory published a short, but highly influential paper 'Conservation of Isotopic Spin and Isotopic Gauge Invariance'. In a series of 22 equations, they outlined how some field theories could be created in which the quanta of the field carried a charge in addition to the particles with which they were to interact. In electrodynamics, the photon carries no electric charge but nevertheless interacts with electrically charged particles. In the case of what became known as Yang–Mills theory, the nuclear force might be due to quanta of the field that also carry the nuclear 'charge' called isotopic spin. The quanta would interact with the protons, neutrons, and also with themselves in a tangled web of strong force interactions.

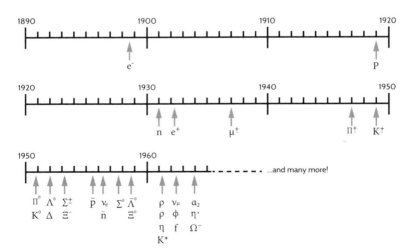

A timeline showing the history of particle discovery.

Murray Gell-Mann was a co-developer of the quark model for nuclear physics.

By the early 1960s, the quark structure of sub-nuclear particles had been postulated by Murray Gell-Mann (1929–2019) and George Zweig (1937–) in an effort to simplify the classification scheme for the burgeoning number of such particles discovered every year. They adopted a mathematical scheme called SU(3), which identifies the group of symmetries that a particular entity, such as a quark, obeys under a set of mathematical operations. Physicists had recognized the relationship between symmetry and quantities that were conserved since 1915, when Emmy Noether (1882–1935) proposed this mathematical relationship from her work in general relativity. For example, the equation for the total energy of a particle remains unchanged if you change the time variable from t to $-t$, which means that energy is conserved. Similarly, for the momentum of a particle, if you change the space variable from x to $-x$, the momentum remains unchanged and conserved. An entire mathematical discipline called Group Theory investigates the geometric symmetries in everything from rock crystals to pure algebraic constructs, largely developed by Evariste Galois (1811–1832). When the Noether Theorem is applied to investigating the conserved isotopic spin of the quarks, this led Gell-Mann and Zweig to a symmetry pattern that can be characterized by a specific symmetry group called the Special Unitary Group of order 3 or SU(3).

EMMY NOETHER
Emmy Noether (1882–1935) was a German mathematician who made major contributions to abstract algebra and

theoretical physics, and was described by Albert Einstein as one of the most important women in the history of mathematics. In 1915, soon after receiving her PhD in mathematics at the University of Erlangen, she was invited by David Hilbert to join the mathematics department at the University of Göttingen. It was there that she was asked to investigate an issue in general relativity, which led her to prove what became known as Noether's Theorem: a conservation law is associated with any differentiable symmetry in a physical system. This became one of the most important theorems ever provided for guiding the advancement of theoretical physics. By 1933, she was expelled from Germany by the Third Reich and became a professor at Bryn Mawr College, but in 1935 she died from complications following ovarian cyst surgery.

Emmy Noether

As a consequence, there would be three quarks with fractional electric charge, but what would be binding quarks together? Certainly not the pions, which would themselves become composite particles made from quarks. Some new quantum particle for the strong force was needed, and it wasn't too long before Yoichiro Nambu (1921–2015) and Moo-Young Han (1934–2016) developed the idea that particles called gluons did the work of binding quarks together. In keeping with the quark SU(3) symmetry, there would have to be eight different gluons to carry combinations of the new 'charge' for the strong interaction called 'colour'. This is because the general SU(N) group has a total of $N^2 - 1$ operations that preserve the symmetry, and for $N = 3$ there would be $N^2-1 = 8$ different gluons whose exchange would preserve the colour conservation symmetry. It wasn't the exchange of the pions that produced the strong nuclear force as had earlier been thought, but instead it was the massless, photon-like gluons that performed this function.

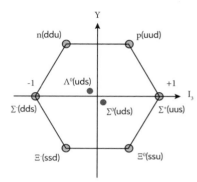

The family of baryons related to the neutron (n) and proton (p) formed from the up (u), down (d) and strange(s) quarks based on SU(3) symmetry.

With the application of SU(3)-based, Yang–Mills field theory, the strong interaction mediated by gluons evolved rapidly into a self-consistent quantum field theory along much the same lines as QED, so that by the early 1970s, the theory and its methods for computation was christened Quantum Chromo Dynamics (QCD) by Murray Gell-Mann. One of the major reasons why the Yang–Mills theory based upon SU(N) symmetries was accepted much as QED had been, was because in 1972 Gerard 't Hooft (1946–) proved that all such theories are in fact renormalizable. This removed the last barrier in accepting this kind of quantum field theory as a basis for describing the entire ensemble of non-gravitational forces known by the 1970s.

Gerard 't Hooft proved that QCD was a renormalizable theory in the same way that QED was, which was a major technical proof of their relationship.

QCD is not a theory for the faint of heart.

Recall that QED populated the vacuum with a rich brew of hidden electromagnetic phenomena among virtual particles such as photons, electrons and positrons. This level of vacuum structure is a gift of the quantized electromagnetic field and its private energy transactions permitted by the Heisenberg Uncertainty Principle. But into this same vacuum we may now add all of the virtual processes contributed by the quark–gluon fields. This is a second level to the vacuum inextricably coupled to the electromagnetic vacuum by the fact that the quarks carry electromagnetic charge (+2/3 and -1/3), and can therefore experience the effects of the electromagnetic vacuum state. Moreover, photons of adequately high energy can decide to spend part of their fleeting lives as virtual quark-antiquark pairs whose interactions are covered by QCD field theory.

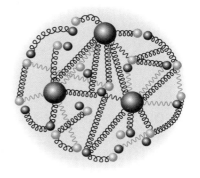

It is possible to isolate the effects of the QED vacuum from those of the QCD vacuum by considering only certain kinds of phenomena such as the anomalous magnetic moment (g - 2) of the electron, which is determined almost entirely by electromagnetic interactions. The QED calculations give a value of 0.001159652181643, which agrees to 1 part per

Inside a proton, the three quarks (large balls) interact with each other by exchanging gluons, but quark-antiquark pairs (smaller balls) can also appear out of the vacuum and interact with all the other components as well.

trillion with the experimental result a_e = 0.00115965218073. By comparison, the strong and weak contributions to a_e are small with values of 0.00000000000167 and 0.000000000000036 respectively.

The heavy brother to the electron, called the μ meson or muon, also has such an anomalous magnetic moment, but in order for the calculations for its value to come out right, allowance has to be made of the QCD vacuum. The QED contribution is calculated to be a_μ = 0.00116584718113 and the QCD contribution is 0.00000000154. The measured value as of a_μ = 0.0011659209.

The tau lepton was discovered in 1975 and is even heavier than the muon with a mass of 1.77 GeV, making it an even better probe of the QCD vacuum. The anomalous magnetic moment of the tau lepton was calculated by Mark Samuel and Guowen Li at Oklahoma State University and Roberto Mendel at the University of Western Ontario in 1991. They found that of the total value of a_μ = 0.0011773, QED amounts to 0.00117319 and QCD contributes 0.00035. The QCD correction is 227,000 times stronger than for the muon. Unfortunately, the best observational limit as of 2011 is that the anomalous magnetic moment of the tauon is between -0.007 and +0.005, so we will have to wait a long time before we can confirm the predicted value and its QCD contribution.

PREDICTING PARTICLE MASSES

The most direct test of QCD would be the actual calculation of the masses of the nuclear particles themselves. These calculations have been fraught with difficulty, not because QCD doesn't tell the theoretician how to make them by adding the contributions from innumerable Feynman-like diagrams, but because no computer on earth today seems to be up to the challenge. In QED, the dazzling accuracy of the predictions of the Lamb Shift could be made by computing the contributions from a few thousand Feynman diagrams. Many more of these Feynman diagrams could be completely ignored because of their rapidly declining contributions to the overall process. Because the strong force is nearly 100 times stronger than the electromagnetic force, no such shortcut seems possible. To achieve the same precision as QED, QCD forces the theoretician to confront literally millions of diagrams, each representing more and more gluons being exchanged among themselves, and the quarks. Just as theoreticians in the 1940s had to resort to the theory of renormalization to expedite finite calculations in QED, theoreticians investigating QCD have resorted to a procedure called Lattice Gauge Theory to try to find a computational shortcut that gives finite results. Even so, the results are far from spectacular compared to QED.

Lattice Theory approximates the subnuclear world as a lattice in space and time. Quarks are free to occupy the nodes of the lattice, while gluon currents travel the narrow roadways of the rungs in the lattice. Currently, lattice spacings of 0.06×10^{-13} cm (or 0.06 fermis) and overall lattice sizes of 2.5 fermis are at the cutting edge of the art. These are just big enough to encompass a modest-sized nucleus. The key difficulty is that the outcome of the calculations depends in part on the size of the lattice you start with. If this tendency continues as finer and finer lattices are used, it may show that *no* shortcuts are possible and only a full, exact calculation will give a unique answer.

Calculations of the masses of protons, neutrons and other strongly interacting particles called baryons have progressed to the point where calculated masses are now within a few percent of the actual masses for these particles using only QCD. The results verify that only a small part of the baryon masses, about 1%, comes from the masses of the quarks themselves. The remaining mass comes from the energy that the quarks carry by virtue of being bound together or confined by gluons within a baryon. The calculated mass of the nucleon (protons or neutrons) is 936 MeV with an uncertainty of 22 MeV, while the known mass of the proton and neutron are 938 and 940 MeV respectively.

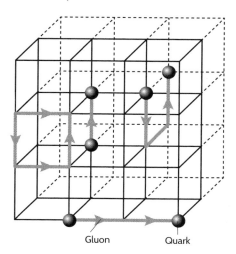

An example of a quark-gluon lattice used to calculate the masses of baryons and mesons in QCD theory.

Enrico Fermi proposed that the weak interaction was a true force in nature like electromagnetism.

THE WEAK INTERACTION

Although the strong interaction is rather intuitive because it can still be thought of as a force that binds things together like gravity, the weak interaction seems to be an oddball force that doesn't produce a push or pull at all. Its effects were first discovered by Henri Becquerel (1852–1908) in 1896. It wasn't really recognized as a force at all until 1933, when Enrico Fermi (1901–1954) used some of the same techniques developed by Dirac to formulate a quantum field theory for nuclear decays.

The weak force causes some particles such as neutrons and p mesons to decay into other, lighter and more stable particles with the consequent emission of a slippery new particle called the neutrino. Fermi proposed that the weak interaction was mediated by a particle called the Intermediate Boson, which at one time or another has also been called the Uxyl, the Schizon or the W, X or Z particle. Following Yukawa's method for estimating its mass, this particle would have to be nearly 100 times heavier than the proton, and in Julian Schwinger's original scheme there would have to be two of them. By 1960 a third neutral particle was proposed by Sheldon Glashow, and so the Intermediate Bosons now consisted of the W^+, the W^- and the Z^0 particles each having the same quantum spin as the photon, but extremely massive.

SPONTANEOUS SYMMETRY BREAKING

There had been several attempts to combine the electromagnetic and weak forces into one theory by 1960. The construction of one toy model after another occupied the time of many theoreticians until 1967, when physicists announced the discovery of a dazzling new, self-consistent 'electro-weak' theory. Steven Weinberg (1933–), Sheldon Glashow (1932–) and Abdus Salam (1926–1996) found that through the use of Yang–Mills theory, and a peculiar mechanism called Spontaneous Symmetry Breaking introduced in 1964 by Peter Higgs (1929–), they could create a theory that combined QED and the weak interaction into a single mathematical framework now called Electroweak Theory. They predicted that there should be three particles to mediate the force (W^+, W^-, Z°) and that at energies above 100–300 GeV the electromagnetic and weak interactions should blend together and become effectively indistinguishable.

The initial response of the physics community was a deafening silence.

A citation study of this paper shows that it received only one reference by 1970; by 1972, 64 papers cited it; and in 1973, 163. Some of the citations were by Weinberg himself, who in 1971 called his own theory 'repulsive'. What happened in 1972 to light a fire under this theory was that Gerard 't Hooft at Utrecht University proved an important mathematical theorem relating to theories such as Weinberg's. The peculiar symmetry-breaking mechanism that was at the heart of Weinberg's theory would *not* shatter the ability of such theories to yield finite answers. Theoreticians were now free to use these theories of spontaneously broken symmetry and did not have to worry that they would be plagued by the same infinities that bedevilled QED 25 years earlier. By 1984, some five years after Weinberg, Salam and Glashow received the Nobel Prize, Carlo Rubbia (1934–) and Simon van der Meer (1925–2011) would receive their own Nobel Prizes for the discovery at CERN of the W and Z bosons.

HIGGS BOSONS

The key ingredient to making electroweak unification a reality involved the prediction, not just of the existence of the W and Z particles, but of a completely new family of particles called the

Higgs bosons. Higgs bosons would be the culprits responsible for breaking the otherwise perfect symmetry between electromagnetic forces and the weak forces. They would do so by causing the quarks, leptons and the carriers of the weak force to gain mass. The introduction of Higgs bosons and spontaneous symmetry breaking into an otherwise clean and compact mathematical theory seemed very ad hoc to many theoreticians.

In his book *The God Particle* (1993) Leon Lederman (1922–2018), writing with Dick Teresi, presented a fast-paced account of the role the Higgs boson plays in the scheme of physics today. The Higgs boson 'is a wraithlike presence throughout the universe that is keeping us from understanding the true nature of matter... The invisible barrier that keeps us from knowing the truth is called the Higgs field. Its icy tentacles reach into every corner of the universe, and its scientific and philosophical implications raise large goose bumps on the skin of a physicist... I have nicknamed [the Higgs boson] the God Particle... because my publisher would not let us call it the Goddamn Particle.' As Lederman points out, however, even the theoreticians who did so much to create this God Particle field mathematically – Peter Higgs, Martinus Veltman (1931–) and Sheldon Glashow – would just as soon distance themselves from the concept. Veltman refers to it as 'a rug under which we sweep our ignorance.' Glashow calls it 'a toilet in which we flush away the inconsistencies of our present theories,' and even Peter Higgs is an unwilling namesake for the particle. Glashow admits that 'I am not at all sure that the Higgs mechanism is nature's choice. It seems to me to be an ugly and contrived construction superimposed upon an otherwise elegant theory. Its virtues are simply that it works, and no one has come up with a plausible alternative – so far.' The much sought-after Higgs boson was ultimately discovered in 2012 at the Large Hadron Collider at a mass of 125 GeV, which made any protestation of its legitimacy a rather moot point. By the present time, electroweak and QCD theory form the two apparently rock-solid pillars of what is now called the Standard Model.

The Large Hadron Collider in Geneva, Switzerland is the most powerful 'atom smasher' of its kind in use today.

THE STANDARD MODEL

The Standard Model consists of six elementary spin-½ leptons and six elementary spin-½ quarks that form the fundamental fermions along with their 12 anti-particles. There are also 12 force-mediating bosons with spin-1: the photon, the 8 gluons and the three intermediate vector bosons. Finally, there is the spin-0 Higgs boson. All of these particles obtain their masses from interactions with the Higgs boson, with an amount of mass-gain that depends on how strongly they interact with the Higgs boson. That means there are 12 adjustable constants for the fermions and 12 for the bosons, making 24 parameters that have to be adjusted along with the strengths of the interactions of the 12 force-mediating bosons with each fermion. That makes for a bewildering number of adjustable constants for this theory which have to be experimentally determined. In addition to these constants, there are a number of other nagging issues that remain, which are not yet easily explained by the Standard Model.

For instance, the Standard Model requires that fundamental particles like the quarks and leptons must be absolutely point-like. It requires that there be three generations of leptons and quarks without any reason for why this must be so. It requires that there exists an all-pervading Higgs field to ensure that the various symmetries in the electroweak theory are broken at low energy so that the world appears as it does to us with electromagnetic forces clearly different from the weak force and certain particles having non-zero rest mass. Finally, it doesn't qualify as a true unification theory because the strong and electroweak forces are thrown together in a rough manner, each with its own separate field theory and calculational techniques.

QED describes the electron as a point-like object because that is the only description for its structure consistent with both the principles of special relativity and quantum mechanics. Any other description involving the electron as an extended object would wreak havoc with this finely tuned and phenomenally successful theory, and lead to glaring discrepancies between predictions and experiments. But theoretical injunctions have never impressed the experimenter. What is the smallest thing

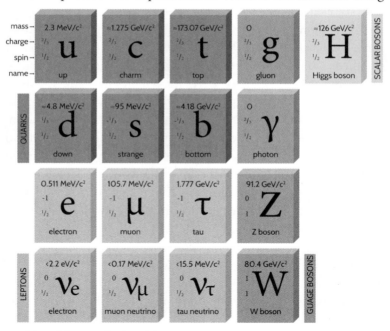

Particles of the Standard Model.

in nature? The electron must certainly win this competition hands down.

The most recent estimate for the size of an electron was provided by Nobel Laureate Hans Dehmelt (1922–2017) at the University of Washington, in which a quantity called the electron gyromagnetic ratio was measured experimentally and compared with the most recent QED calculation of it. This physical quantity occurs because the electron possesses spin and is charged. The movement of charges causes magnetic fields, so you can think of the electron as a tiny magnet with a north and south end. The gyromagnetic ratio just measures how strong its magnetic field is given that it has a ½-unit of quantum spin. The measured value was 1.001159652188 with an uncertainty of 4 parts per trillion. The value mathematically predicted assuming a point-like shape for the electron was 1.001159652133 with an uncertainty of 29 parts per trillion. The discrepancy between the two was deemed significant and might mean that at a scale of 10^{-20} cm, the QED model and experiment are beginning to differ. Dehmelt argued that this is a sign that the electron might be extended in space with about this radius. Then again, a conservative interpretation would be that even at this minute scale, the electron still has no significant extension in space.

In any event, the bottom line is that so long as we restrict our experiments and theoretical predictions to regions larger than 10^{-20} centimetres, the electron looks just like it always has: a mathematical point in space. This length scale also means that for collisions or other interactions with energies less than 10 million GeV the electron is still point-like and you can go right ahead and use QED to describe its interactions. Current 'atom smashers' such as the Large Hadron Collider at CERN routinely reach energies up to 13,000 GeV so that the electron will still look point-like for the foreseeable future.

The electron is the least massive particle we identify as matter, but what does a completely massless particle look like? The photon is probably the closest nature gets to this ideal state. Not even the neutrino can make this claim unambiguously. Quantum

Electrodynamics requires that the photon must be absolutely without any mass, apart from whatever effective mass it might have as a consequence of its energy and Einstein's relation $E = mc^2$. An early measurement of the photon's mass was described by Alfred Goldhaber and Michael Nieto (1940–2013) in 1971. They used a variety of electrodynamic phenomena to place stringent constraints on just how massive a photon could be without violating any known experimental result. If the photon contains as much as 2×10^{-47} grams of mass (10^{-14} eV), departures from Coulomb's inverse square law for electrostatics would be detectable under laboratory conditions by a reasonably careful undergraduate student in a college physics laboratory. Even the shape of earth's magnetic field would be measurably different if the photon were as massive as 10^{-48} grams.

GRAND UNIFICATION THEORY

Theoreticians are an impatient lot. Flushed with the excitement of electroweak unification, the hunt began in earnest for a more unified theory combining QCD and Electroweak theory. It is precisely at this point that any attempt at describing the development of this idea gets immediately bogged down in an unrelenting parade of difficult terminology. Arguably, the description of quantum physics up to this point has benefited from the introduction of a few well-worn terms derived from experimental investigations such as quarks, fields and symmetry. But now the focus turns away from experimental work and explores the purely mathematical investigations of a seemingly endless set of possibilities for grand unification theory; commonly referred to by its acronym GUT. Virtually the entire discussion hinges on the exploration of symmetries identified by their nomenclatures.

Although the Standard Model used the individual symmetry groups denoted by SU(3), SU(2) and U(1) to describe the strong, weak and electromagnetic interactions, an even larger mathematical group was sought as a unifier of these three forces, and the question was: Which one to use? The mathematician Elie Cartan (1869–1951) had long since systematically classified all of the possible 'simple' symmetry groups into the only eight possible categories: SU(N), SO(N), Sp(N), G2, F4, E6, E7 and E8. The process of finding a symmetry group large enough to accommodate the known fields and particles in the Standard Model; to reproduce the electroweak symmetry breaking at low energy; to yield a single interaction strength; and at the same time to account for the known families of particles, is a challenging exercise. The only symmetry groups that appeared to have the correct attributes are identified by the labels SU(5), SO(10) and E6. SU(5) was the first one investigated by Howard Georgi and Sheldon Glashow, E6 by Y. Achiman and B. Stech, and SO(10) by Howard Georgi and Dimitri Nanopoulos. Other possibilities such as SU(3) × SU(3) have also been considered, but have their fill of problems, especially in keeping down the number of new particles they require to make them work. Mark Srednicki (1955–) at the University of California Santa Barbara once commented: 'It is impossible to get this pile of junk to come out anything like the Standard Model.'

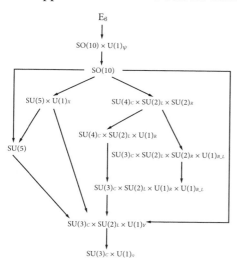

The complex path by which E(6) symmetry is broken into smaller symmetry groups including SO(10), SU(5), and the Standard Model groups: SU(3), SU(2) and U(1).

Georgi's SU(5) model enjoyed quite an exciting birth, but in the 20 years since it was first announced it began to show its age. By 1984 it was deemed an exciting first go at finding a GUT, but seriously incomplete. Among its faults were the multiplicity of apparently ad hoc Higgs fields and the prediction of proton decay. SU(5) is still being tested experimentally today in searches for proton decay and neutrino mass, but theoretical interest in this area seems to have led to a dead end. Today, it is widely recognized that SU(5) with its minimal compliment of 25 Higgs fields suffers from too many technical problems to be the correct grand unification theory (called GUT).

Although there is no single GUT that is universally agreed upon, the contenders do have several common and significant predictions to make about the physical world. Two critical temperatures emerge, signifying the onset of two major symmetry breaking phases. Much as ice turns to water at 0° C and then to steam at 100° C, the laws governing matter and its interactions change abruptly at the temperatures characterizing the Electroweak and GUT unifications. The nature and number of distinguishable forces change from the present 3, to 2 at the Electroweak temperature and then to only 1 at the GUT temperature much as water cools to a liquid and then crystalizes to ice. The temperatures at which these crystallizations occur are truly fantastic. The Electroweak transition is predicted to occur near 200 GeV, whereas the GUT transition occurs at 1000 trillion GeV.

At the same time that these theories predict unification, they also require the existence of new families of particles. Were it not for these ephemeral, yet to be discovered, particles, ponderable matter as we know it would not exist and all particles in the universe would be massless like the photon. GUTs also predict that new families of Higgs bosons should exist, but they are far more massive and numerous, and the theories provide few constraints on their physical properties and self-interactions. Nothing is known about how many Higgs particles there are, what determines their number, or how they interact among themselves let alone with the fermions. The action of the GUT Higgs bosons also leads to a revision of our understanding of the vacuum state, this time with spectacular physical consequences. Among these are a whole host of cosmological phenomena.

Key Points

- Yukawa Theory uses Heisenberg's Uncertainty Principle to relate the range of a force to the mass of the particle that mediates it during its exchange between two particles.

- The strong interaction is produced by the exchange of gluons between the constituent quarks, which constitute protons, neutrons and other baryonic particles.

- The weak interaction involves the exchange of massive W and Z bosons and causes quarks to change their identities, resulting in the decay of baryons into lighter particles.

- Symmetry breaking causes one force such as the weak force to resemble another force such as electromagnetism when a specific energy threshold (temperature) is reached, similar to the phase change in water as it reaches its freezing or boiling point.

- Higgs bosons interact with fermions and bosons to cause them to change their masses, which can result in one force resembling another when the mediating particles are involved.

- The Standard Model is the term used to describe the mathematics of the 12 bosons responsible for the strong, weak and electromagnetic forces, and the 12 matter particles called fermions, along with the Higgs boson, together with all of their anti-particles.

- Grand unification theory is the model for unifying the diverse particles and fields of the Standard Model into a single, naturally unified description that would underly QED, QCD and Electroweak theory.

CHAPTER 11
Supersymmetry

Electroweak theory reveals a perfectly unified world in which the electromagnetic and weak forces have the same strength and all quarks and leptons are massless. This is not the world we live in. Theoreticians have had to damage this perfect mathematical world by introducing the Higgs boson as a way to break its symmetric condition into the imperfect reality we experience where electromagnetic and weak forces behave very differently. Symmetry and broken symmetry are all around us if we know where to look. Because symmetry plays such an important role in modern quantum field theories, let's explore this in further detail.

It is obvious that symmetry in time is not perfectly obeyed; after all, we do get born, age, and eventually die. We sense that time runs in only one direction so that we never remember events in our lives that are to happen in the future. Yet, the equations that describe the subatomic world work just as well if time were running backwards rather than forwards. Nature clearly enjoys flirting with symmetry, but does not slavishly obey its every geometric mandate on form and function. There are two phenomena that have historically served as the archetypes for how the Higgs field works its magic: magnetism and superconductivity.

An ordinary bar magnet heated to high temperature but still maintaining its original shape, will no longer have a recognizable north or south pole. The individual magnetic domains of the atoms in the bar become randomly oriented as the atoms twist and move about. If you happened to be a well-insulated, bacterium-sized physicist living in this iron bar, you would see that no matter where you moved, things looked pretty much the same. There

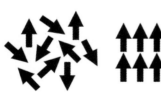

Applied Magnetic
Field Absent

Applied Magnetic
Field Present

Above the Curie temperature, magnetic domains remain disordered in the presence of a magnetic field (left), but below this temperature they become permanently ordered (right).

would exist for you a certain symmetry in your world called displacement symmetry, such that if you shifted any two atoms from their original locations, you could not tell which had been located where, because they would blend into their new locations as well as they had into their previous ones. As the temperature of the bar is lowered below a temperature of 770° C, called the Curie temperature, the mobility of the atoms begins to decrease in such a way that their collective energies are always as low as possible. When one atom feels its north end surrounded by too many other north ends from neighbouring atoms, it will flip over to minimize the magnetic stresses that are being exerted upon it. You can feel these stresses yourself by trying to touch the north poles of two magnets together. This readjustment continues until the bar is cold to the touch, but as a consequence of this cooling, a radical change will occur in the large-scale properties of the iron bar. The minimum energy occurs when a particular end is either a north-type pole or a south-type pole. Now as you moved around within this iron bar, you would see that the previous state of symmetry where things looked pretty much the same everywhere no longer holds true. In some regions of your space, a force appeared that pointed north while in a more distant region, the same force permeated your space but pointed south. If you moved an atom from one end of your universe to the other, it wouldn't blend into the rest of the scenery but stick out like a sore thumb with its poles pointed the wrong way. You would conclude

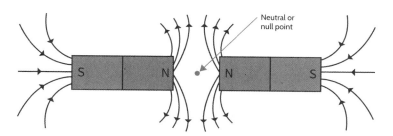

The magnetic fields of bar magnets provide an analogue to how symmetry-breaking events imply the existence of other fields in nature.

that, somehow, the local space displacement symmetry had been broken by the simple act of cooling the bar of iron.

Imagine that you had one set of equations describing the iron bar when it was not magnetized. You would need a completely different set to describe the properties of the bar when it was magnetized. Never mind what these equations are; just think of them as a set of tables which predict how the observable quantities such as density, temperature, magnetic field strength and orientation are distributed within the iron bar. How would you go about trying to unify these two descriptions? After all, you are talking about only one physical system. You might think that there should be only one self-consistent mathematical theory to describe it. The clue to finding such a theory can be found by following what happens at different temperatures.

You know that the iron bar consists of individual atoms, so the appearance of the large-scale field and the simultaneous breaking of displacement symmetry must somehow have to do with the orientations of large numbers of these atoms lining up in a certain way. So you create a mathematical model showing how this works as you change the temperature of the bar. The model seems to work exceedingly well, but it reveals something rather strange about the space you live in. The appearance of the magnetic field adds a quantity of energy to what you used to think was simply empty space. When the bar is cold, all of your experiments have to take into account this new 'vacuum energy', but in addition to this is yet another strange property: The state of the vacuum with this lowest energy is actually *two* states, which you call North-type or South-type. It takes a long time for you to convince your colleagues that the nothingness of empty space can not only have an energy to it but can come in two flavours.

So far we have talked about magnetism and how phase transitions emerge when some systems are cooled below a critical temperature. What does this all have to do with fundamental particles and the nature of mass?

A lot.

Yoichiro Nambu developed the idea that symmetries are broken when new particles appear.

By 1911, it was known that certain materials such as mercury lost their internal resistance and became superconductors when subjected to liquid helium temperatures near 4 kelvins. For the next 50 years, a steady refinement in the quantum mechanical explanation for superconductivity continued, leading to the concept of Cooper Pairs; pairs of electrons that acted to create the superconductivity effect at low temperature.

Because Cooper Pairs behave like connected pairs of electrons, in some sense they act like particles interacting to lower the conduction energy of a metal so that it becomes a superconductor. In 1961, Yoichiro Nambu (1921–2015) proposed that the appearance of these Cooper Pair 'pseudo-particles' in the prevailing theory of superconductivity could be the analogy he was looking for to explain how symmetries are broken among the various fields that littered the landscape of particle physics. Between 1960 and '61, Abdus Salam (1926–1996) and John Ward (1924–2000) went so far as to propose that the pseudo-particles are not merely pairs of ordinary particles, but were real physical particles. In superconductivity, the Cooper Pairs arise spontaneously from the pre-existing population of electrons in the metal, but what causes pseudo-particles to suddenly make their appearance in quantum field theory and upset the previous symmetry in the system? Physicist Jeffrey Goldstone in 1962 (1933–) went on to conjecture that the vacuum state itself was the ultimate source of these pseudo-particles. The vacuum state contained a new field whose quanta came to be known as Goldstone bosons, and which interacted with elementary particles to break symmetries.

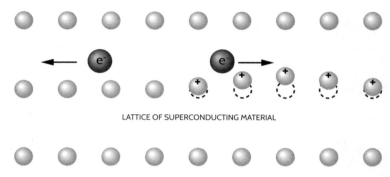

An example of how Cooper Pairs form in a superconductor.

PETER HIGGS

Born in England in 1929, Peter Higgs received his PhD in physics in 1954 at Kings College on problems related to molecular physics. During his time at the University of Edinburgh in the early 1960s, he became interested in the problem of how particles became massive. Building on the work by Yoichiro Nambu and Jeffrey Goldstone, Higgs associated the symmetry-breaking field with a real physical field whose quanta are now called Higgs bosons. Massless particles 'eat' these bosons and become massive in what is now called the Higgs Mechanism. Peter Higgs and François, Baron Englert (1932–) shared the 2013 Nobel Prize in Physics for this work, which had led to the discovery of the Higgs boson at the Large Hadron Collider in Geneva in 2012.

Peter Higgs.

In 1966, Peter Higgs examined a new model for spontaneous symmetry breaking, in which the new massive Goldstone bosons in his theory, now called the Higgs Bosons, would be absorbed by the originally massless spin-1 particles. The force represented by the now massive spin-1 field must behave differently than the force that was originally produced by the massless photon-like field. It could no longer be a long-range force like gravity or electromagnetism, but as Yukawa had previously calculated, the resulting force must now have a range of atomic or even nuclear dimensions. The stage was now set for Steven Weinberg, Abdus Salam and George Glashow's electroweak theory in which the Higgs mechanism was applied to the development of a rigorous and detailed quantum theory of the electromagnetic and weak interactions. There would now be only one kind of theory that could lead to different kinds of particles and fields as the energy of the interactions was lowered below some critical value.

At very high energies above 200 GeV, the photon and the W particles start out life as completely massless particles. In this state, the interaction between these fields and the electrons and neutrinos would behave as though there were an exact symmetry between the forces. In group theory terms, the symmetries obeyed what was called SU(2), which required $N^2 - 1 = 3$ new particles to maintain the symmetry, which were the W^+, W^- and Z^0 bosons. A single Higgs scalar field was also included, but since it was massless in this perfect symmetry state, its emission and absorption by the W and Z particles would not affect their masslessness.

Below an energy of about 200 GeV a surprising thing happens: the Higgs boson becomes heavy and is absorbed by the W and Z bosons, but now the plague of mass is visited upon the W and Z bosons as well. The photons do not gain mass. They are immunized from the plague because the Higgs bosons do not interact with photons. Just as the massive Higgs boson can be absorbed by massless spin-1, W and Z bosons, it can also interact with the other fermions, such as the electrons, neutrinos and quarks, to give some of them their characteristic masses. If you feel somewhat uncomfortable with this plague-like process mediated by the Higgs field, you are in good company. Where and how do the Higgs bosons get their mass? The answer seems to be in a nearly circular line of reasoning which sounds much like the old chicken and egg paradox: which event came first?

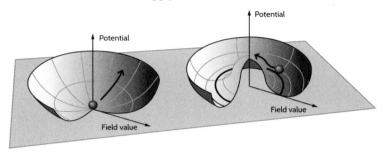

The Higgs potential at high temperature (left) and at low temperature (right).

Originally, the Higgs bosons had no mass above some energy where the minimum energy of the vacuum only had one solution: the solution where the minimum energy of the Higgs field was zero (see figure, left). Under these conditions, the Higgs bosons are exactly massless. After cooling, the vacuum state developed two distinct minimum energy states, but in each one, the Higgs bosons gained mass (see figure, right). The reason that the vacuum state possessed an energy is that the Higgs bosons are capable of interacting with themselves, and it is the behaviour of their self-interaction with

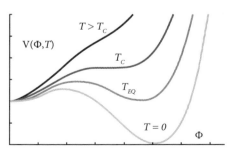

Temperature dependence of Higgs interaction potential energy.

falling temperature that determines the shape of the vacuum energy curve.

What we identify as mass is just another name for how strongly a particle is tied to the Higgs field, and what state this Higgs field is in. The next time you are tempted to use a crash diet, just remember that our masses, which determine our weight, are written not only in the atoms in our bones, but in their coupling to the cosmic Higgs field for which we can do nothing.

The Higgs mechanism and its vindication by the discovery of the Higgs boson in 2012 at the Large Hadron Collider in CERN, has fundamentally changed the way physicists think of the vacuum state. It was one thing to consider a single vacuum populated by virtual photons (QED), virtual Ws and Zs (electroweak) or virtual gluons (QCD), but now this vacuum state has to be indexed by the properties of yet another field in nature, which is quite different than the spin-1 fields of the photon, gluon and W and Z particles. In this new vacuum state indexing determined by the spin-0 Higgs field, there can exist more than one lowest-energy vacuum state. We can think of two separate universes with their own separate but identical QED and QCD physics living on top of each of the two identical but separate Higgs vacuua.

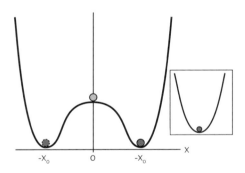

As we cool a bar magnet below its Curie temperature, the randomized north and south poles of the individual magnetic

The Higgs vacuum has two energy minima at low energy and only one, symmetric, minimum at high energy.

domains begin to align and give one end of the bar magnet either a north or south-type polarity because there are two possible minimum-energy solutions available for a specific geometric point on one end of the bar or the other. Similarly, for the Higgs field at low temperature, the particles in the entire universe can find themselves in either one of the minima $(-X_0)$ or the other $(+X_0)$. The resulting mass of the Higgs boson that observers in either universe observes is simply X_0.

The advent of symmetry breaking as a new concept in quantum field theory led to the question as to whether such spontaneously-broken field theories were still renormalizable or not. If not, this marvellous new idea would lead to the reappearance of infinities in calculations in, for example, QED, which had already been made finite through renormalization. In 1971, Gerard 't Hooft at Utrecht University proved that, in fact, there was nothing to be feared by breaking symmetries using the Higgs mechanism. This led to a dramatic upsurge of interest in trying to unify the strong, weak and electromagnetic forces into a larger theory in which the symmetry between the strong and electroweak forces could be broken in the same way.

SUPERSYMMETRY

As we saw in the previous chapter's discussion of GUTs, the 1970s were an intense period of theoretical investigation during which many simplified models for how to unify the three forces were investigated in exhausting mathematical detail. One of the most exciting and far-reaching of these investigations was carried out in 1974 by Julius Wess (1934–2007) at Karlsruhe University, and Bruno Zumino (1923–2014) at CERN. In a pair of papers describing something called supergauge transformations, they presented some of the intricate mathematical work they had been doing for the last few years. These transformations acted upon a set of what were called superfields, which led to integer-spin and half-integer spin particles changing their spin assignments. In other words, they had found a mathematical theory in which 'dogs' could be transformed into 'cats'.

Previously, the kinds of transformations that theoreticians had worked with preserved the spin character of the fields they operated upon: fermions always remained fermions and bosons always remained bosons under the symmetry operations of SU(2) and SU(3) for example. But if you defined a new kind of a field as having a mixture of integer and half-integer components, the field would remain unchanged while its sub-components changed their roles under the Wess–Zumino supergauge transformation, later called supersymmetry. What does this all mean? Let's consider a little analogy between supersymmetry and a pocket full of loose change.

Imagine that you have just completed a tour of several foreign countries and your pockets are heavy with a variety of coinages from Germany, Sweden, England and the United States. The total value of all the currencies is a fixed number and represents a fixed amount of gold. In supersymmetry theory, this represents the magnitude of the superfield. Under a supersymmetry transformation, this superfield can be rotated so that its magnitude remains the same, but the various components of the field must change in a well-defined way. In our coinage analogy, we can go to a bank and there perform a currency transformation on each of the coinages to transform them into one of the other currencies, without changing the total value of our money in terms of its value in gold. The fermion (electron, quark, neutrino) and boson (photon, gluon, W and Z) components of the superfield are the coinages of nature, and can be transformed one into the other via the supersymmetry transformation.

The implications of the Wess–Zumino supersymmetry theory were revolutionary. The fermionic and bosonic particles had to have masses that were strictly related to one another, and even more surprisingly, some of the pesky infinities that plagued the field theories of each of these components individually, went away because of the constraints imposed by the new supersymmetry. The mathematical structure of the superfield is fixed so that only the following spins get paired together: [spin-2, spin ¾] and [spin 1, spin ½]. This means that for every lepton and quark with spin-½, there is paired with it a supersymmetry partner field with spin 1; for every graviton with spin-2, there is a spin ½ 'gravitino' field. But when these new partner fields are added to the calculations of particle dynamics, they miraculously contribute their own infinities which are opposite to the ones contributed by the ordinary matter.

Returning to our coinage analogy, imagine that you wanted to return to the United States and purchase a dinner. Although your currency value in an equivalent amount of gold is more than sufficient to purchase the dinner, it is distributed over four different currencies. So, you have to perform a supersymmetry transformation that rotates each of the coinages into US currency before you can perform the calculation of purchasing your dinner. With supersymmetry theory, a calculation would be performed by converting the other fields in QED and QCD into the [2, ½] gravity superfamily, and then carrying out the calculation in terms of these spin-2, spin ½ field components. This would lead to a finite result, and you then transform the result back into the original superfield components of QED and QCD to get the answer you want for the process you are considering.

A curious feature of the supersymmetry transformation is that when it is performed twice on the same particle, the relationship between the field before and after is one of translation in spacetime – the particle moves from one location to another. Because gravity and spacetime are so intimately linked, this could only mean that buried somewhere inside the mystery of supersymmetry is the hint of a theory that automatically included gravity. By 1975, Bruno Zumino and Sergio Ferrara (1945–) were able to show that supersymmetry implied that a pair of super fields should exist; one of which had a spin-2 and a spin ½ component paired together as a unit, which they later christened the gravitational supermultiplet. This became known as supergravity theory.

For several years afterwards, supersymmetry models were explored with increasing excitement until 1977, when Murray Gell-Mann, at a meeting of the Washington, D.C. American Physical Society, announced some rather troubling news. Even the largest supersymmetry theory known at that time, SO(8), didn't have enough mathematical slots in it to accommodate the known fields in nature. SO(8) had:

1 graviton field
8 spin-³⁄₂ fields
28 spin-1 fields
56 spin-½ fields
70 spin-0 fields.

No one at that time had ever seen a spin-0 field, let alone 70 different kinds of them, and the 56 spin-½ fields had properties that did not match up with the known 6 leptons and 6 quarks or their anti-particles. Could the Higgs fields that are also spin-0 particles be accommodated by SO(8) supergravity? The answer to this hopeful question also seemed to be: No. They would still have to be added as an extra set of fields above and beyond the 70 supersymmetric, spin-0 fields. Supersymmetry would also need to be a broken symmetry in our universe because none of the superpartner fields to the ordinary quarks, leptons and bosons had been observed at energies below 100 GeV.

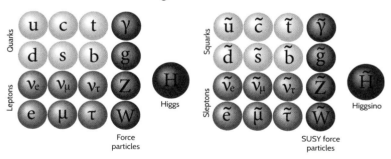

In supersymmetry theory, the normal elementary particles of the Standard Model on the left are paired with new more-massive particles on the right.

HIDDEN DIMENSIONS TO SPACE
In 1976 a simpler way of looking at supergravity, called dimensionally extended supergravity, had appeared. It didn't

resolve all of the problems of the supersymmetry particle mismatches, but it did show that some interesting simplifications were possible in the cumbersome SO(8) theory. In the 1920s, an idea called Kaluza–Klein theory proposed that spacetime was 5-dimensional, and the addition of this new fifth-dimension allowed Einstein's equation for gravity to include a new field in Nature that strongly resembled the electromagnetic field. After laying dormant as an idea for 60 years, in 1982 this idea was resurrected to show that SO(8) in 4 dimensions, called $N = 8$ Supergravity, could be recast as the far-simpler SO(1) theory, but in 11 dimensions. By adding additional dimensions to spacetime beyond the normal four, the mathematics describing these theories, and the complement of fundamental fields, became much simpler in 11 dimensions. When the dimensionally-extended theory was reduced back to a 4-dimensional theory, you could recover the full complement of fields predicted by the more complex SO(8) theory. The exciting thing about $N = 8$ supergravity was that certain kinds of calculations involving gravity would lead to finite results. However, by around 1984, even this theory was abandoned because it didn't completely do away with infinite answers and the particle mismatches, even by using the 'trick' of dimensional reduction.

During the late 1970s an increasing fraction of theoretical attention had become focused on the search for a self-consistent, infinity-free GUT for the strong, weak and electromagnetic forces. Supergravity and supersymmetric GUTs (called SUSY GUTs) were the object of a veritable feeding frenzy of activity as this new motherlode was mined. A bewildering number of ideas were spawned which impacted the theoretical structure of spacetime and the vacuum. New phase transitions were identified at energies between 100 and 10^{15} GeV, implying new patinas of vacuum states littering spacetime. Dimensional extension was applied to supersymmetry theory, but required spacetime to have far more than 4 dimensions; perhaps as many as 11. These additional dimensions would be rolled up into subspaces resembling tiny spheres attached to each point in 4-dimensional spacetime, but other more exotic geometries for these subspaces were also investigated. During the late 1970s and early 1980s, there came a veritable explosion of popular articles as physicists boldly explored these higher-dimensional landscapes. A 1985 *Scientific American* article 'The Hidden Dimensions of Spacetime' written by Daniel Freedman (1939–) and Peter van Nieuwenhuizen (1938–) tried to set the record straight on what all of this meant, but any intuitive understanding of 11-dimensional spacetime seemed to elude the grasp of even the most ardent science popularizers.

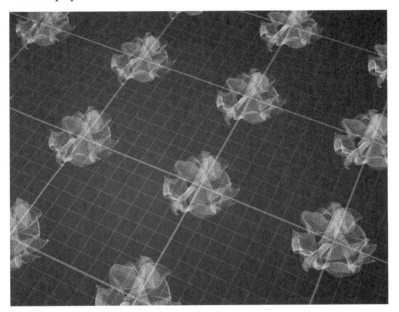

An example of the hidden dimensions for ordinary 4-dimensional spacetime predicted by supersymmetry and supergravity theory.

SUPERSYMMETRY AND THE STANDARD MODEL

The experimental case for the existence of supersymmetry has not been what was expected after nearly 40 years of searching. As we have seen, beginning in the 1930s, physicists developed a detailed set of mathematical models of four fundamental forces in the universe: gravitation, electromagnetism, and the strong and weak nuclear forces. The principles of quantum field theory described in Chapter 9 show that the three basic non-gravitational 'force fields' are themselves constructed from the exchanges between new families of particles and the elementary electrons and quarks. The photon transmits electromagnetic forces; eight different gluons transmit the strong nuclear force that holds quarks together inside protons; three intermediate vector bosons transmit the weak nuclear force, which is responsible for radioactivity; and some physicists believe that gravitons transmit gravitational forces, but these particles remain undiscovered.

In the Standard Model, physicists group the description of matter and forces by these various detailed mathematical theories. Adding supersymmetry to the Standard Model doubles the number of fundamental particles, and this is called the Minimal Supersymmetric Standard Model (MSSM). Not only does MSSM correct a number of problems with the Standard Model, but it also introduces a new candidate particle for the mysterious material that makes up about 24 per cent of the universe's gravitating mass – dark matter. The Standard Model has no way to explain this invisible mass, but the least massive particle of MSSM, called the neutralino, has exactly the right properties to account for dark matter.

Some physicists look at MSSM as the most well-developed and easiest way to move beyond the Standard Model without drastically adding more assumptions. The theory also provides new predictions that can be tested at the world's most powerful particle accelerator: the Large Hadron Collider (LHC). The known superpartners to Standard Model particles should have masses of a few teraelectron volts (TeV), which the LHC can easily access across its energy range from 0 to 13 TeV. The LHC began operation in 2008, and in 2012 discovered the Higgs boson, with a mass of 125 GeV. This was a crucial particle that had been sought for over 50 years in order to validate the electro-weak portion of the Standard Model. By 2019, many other predictions of the Standard Model have been tested and experimentally verified with the help of the LHC. However, what has been resounding good news for the Standard Model has been very bad news for any signs of new physics required for supersymmetry and the MSSM. The study of the decay of exotic particles called B-mesons has turned up no sign of supersymmetry. These particles should have decayed far more quickly with the help of the new family of massive particles provided by supersymmetry, but scientists have found no indication for this effect among the trillions of decays studied. More strikingly, there has been no evidence for the supersymmetric particles themselves below 7 TeV, which is interpreted as a major failing for the simplest supersymmetry models (and specifically MSSM).

The stakes are extremely high: If supersymmetry is found, the discovery road will be opened to explore theories beyond MSSM. If they continue to find no evidence of supersymmetry at the LHC, then the simplest versions of MSSM must fall, and the entire 50-year exploration of this elegant symmetry in nature may also come to a halt along with confidence in any theories that rely on supersymmetry. The most exciting of these Theories of Everything for the last half-century has been superstring theory.

Key Points

- Symmetry-breaking can be understood by analogy with magnetism as a magnet is cooled below its Curie temperature.

- The potential energy of space due to the Higgs field takes on two different forms, depending on whether space is above or below a critical temperature related to the mass of the Higgs boson.

- Supersymmetry is a new symmetry in nature which combines fermions and bosons into new families that allow for the mathematics unifying the forces and particles of nature.

- Supersymmetry predicts that new massive particles should exist for each known particle.

- Currently, despite intensive searches at the expected energies of these new particles, evidence for supersymmetry has yet to be found in the expected places.

CHAPTER 12
String Theory

At the time the quark model was being developed in the early 1960s, another idea of how quarks were bound together inside nuclei was developed to explain the regular mass spacings in mesons that were modelled as pairs of quarks. The string represented an unknown binding force with the quarks located at each end of the string. This theoretical approach was developed into what was called Bosonic String Theory in 1974 by John Schwarz (1941–) and Joël Scherk (1946–1980), in which each vibrational mode of the 1-dimensional string represented a different particle. However, a self-consistent theory of the bosonic string led to the realization that it would work only within a 26-dimensional spacetime framework; it permitted faster-than-light tachyon states; and, moreover, it did not include fermions like quarks and electrons.

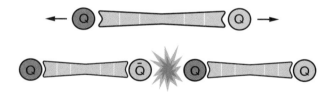

A quark string contains two quarks on either end of a 'string', which when pulled too tightly can snap to produce a new pair of quark strings, but not a single quark by itself.

JOHN SCHWARZ

John Schwarz was born in 1941 and received his PhD in physics in 1966 at the University of California, Berkeley. His thesis advisor was Geoffrey Chew (1924–2019), whose work in earlier theories of proton structure and the strong interaction made him a pioneer in what is called S-matrix theory. This was a string-like interpretation of particle (hadronic) structure, which

John Schwarz.

later influenced Schwarz to continue work in hadronic string theory, leading to his partnership with Michael Green (1946–) in 1984 to develop the basis for what is now called superstring theory. Since 1972 he has been at the California Institute of Technology. In 2014 he and Green shared the Breakthrough Prize in Fundamental Physics, often called the 21st-century Nobel Prize.

In 1984, John Schwarz and Michael Green announced Superstring Theory. Henceforth, particles would not be thought of as point-like concentrations of energy, but as 1-dimensional, vibrating strings of energy. Both bosons and fermions could be accommodated in this 'superstring' theory and represented as different classes of vibrations of these strings in 4-dimensional spacetime. Particle world lines would be fattened from spaghetti-like, 1-dimensional tracks in spacetime, to macaroni-like tubes, and with this new structure all infinities would vanish without any need at all for renormalization. The only problem is that: 1) spacetime would have to have either 10 or 26 dimensions in order that the theory be self-consistent; and 2) the theory would naturally work only at energies near 10^{19} GeV. The lowest mode of string oscillation would yield particles with no rest mass at all. The next-highest mass range would be 10^{19} GeV.

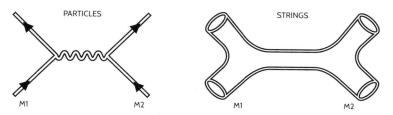

Instead of a point particle, particles were closed strings that moved through spacetime to form 2-dimensional surfaces. The vibration of these surfaces represented different particles.

Superstrings would come in two types: open strings with end points, and closed loops. The former types would be anchored in 4-dimensional spacetime and form the basis for the Standard Model particles while the closed loops would represent the individual gravitons that produce the gravitational force. Following what became known as the First String Revolution in 1984, investigators found four other types of superstring theory less than one year after the initial string theory announcement. The full collection of theories now included Type I, Type IIA, and Type IIB discovered by Schwarz and Green in 1982, and the so-called closed string Heterotic theories SO(32) and E7 × E7. Each theory was within itself self-consistent, complete and described the physics and phenomenology of particle states as excitations of strings vibrating in a 10-dimensional spacetime. In order to represent the physical world, four of these dimensions were the normal 4-dimensional spacetime of relativity, but the additional six dimensions were accessible to the strings only and formed a specific geometric configuration. These geometric configurations had to have a specific topology in order that they provided the

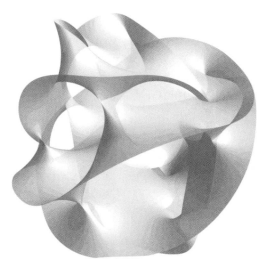

Example of the geometry of a Calabi–Yau space.

internal symmetries needed for the strings to manifest the known particle states. These compact spaces of subatomic size and finite extent were called Calabi–Yau Manifolds.

By the early 1990s, there was considerable frustration in the string theory community. String theory was considered to be a prototype 'Theory of Everything' that unified the Standard Model particles and fields with gravity, but which of the five string theories was the correct theory? In 1995, Edward Witten proposed that the many technical similarities between the five string theories could be explained if there were just one theory operating in 11 dimensions called M-theory. In essence, with this extra dimension added, a new symmetry could be added which allowed five different projections of the larger theory to form the five separate string theories. This is much like rotating a cube in 3-dimensional space to reveal the six individual 2-D faces of the cube. The M in M-theory could be interpreted as 'magic', 'mystery' or 'membrane' depending on the perspective of the physicist. This period, which started in 1995, is often called the Second String Revolution, and led to a number of interesting proposals by Lisa Randall (1962–) and Raman Sundrum (1964–) as to the nature and dynamics of objects in M-Theory called branes.

The language of M-theory involves terms like 'the Bulk' to represent the full 11-dimensional spacetime. It also focuses attention on 4-dimensional subspaces called branes (after 'membrane'). Our spacetime is one of these braneworlds in which open-ended strings are anchored at their endpoints and from their properties form the Standard Model particles and fields we know. The closed loops represent gravitons, and so gravity is a force that can operate across the 11-dimensional Bulk, unlike Standard Model particles. Also, the weakness of gravity is accounted for because only a portion of it is present in any 4-dimensional braneworld. Because Standard Model particles do not extend as loops more than a few Planck lengths (10^{-33} cm) beyond each braneworld, it is not possible for us to detect other braneworlds in the Bulk.

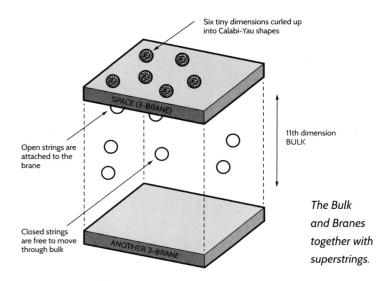

Six tiny dimensions curled up into Calabi-Yau shapes

SPACE (3-BRANE)

Open strings are attached to the brane

11th dimension BULK

Closed strings are free to move through bulk

ANOTHER 3-BRANE

The Bulk and Branes together with superstrings.

The most disturbing feature of superstring theory is that, depending on the geometry of the Calabi–Yau manifolds, you get a different Standard Model. By some estimates, there are as many as $10^{272,000}$ of these possibilities. Each of these leads to its own model for the quantum vacuum, the cosmological constant, and particle and field assignments. The ensemble of string theory states is called the Landscape, and the key problem is to understand how our particular physical world and Standard Model were selected from among the Landscape. The sum total of all these possibilities is also called the Multiverse. In some sense, this is akin to the description of quantum states as points within a Hilbert Space of possibilities. Only a specific observation collapses this space of possibilities into a single quantum state that is actually observed. For string theory, some version of an Anthropic Principle has been proposed as a boundary condition. In other words, our very existence as observers reduces the Landscape possibilities to only a very small subset of possibilities consistent with the world we see. The role of observer participation has been an intriguing solution to many quantum mechanical problems since John Wheeler proposed this idea in 1994: 'No phenomenon is a real phenomenon until it is an observed phenomenon... we are participants in bringing into being not only the near and here, but the far away and long ago.' We have already touched upon many of these ideas in Chapter 5, when discussing the act of measurement from a quantum mechanical perspective.

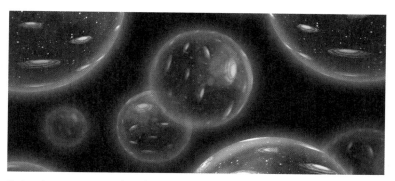

String theory proposes that there are millions of different, self-consistent, theories of the universe and that each one may in fact represent the possibilities for actual universes - an idea known as Multiverse Theory.

Throughout the Second String Revolution to the current time, string theorists have wrestled with forcing this 'beautiful' theory to make predictions about particle physics in our low-energy world at the bottom of the energy ladder. Many feel that nature has been unkind to us by providing physicists with a theory from the 22nd century, while giving us only mathematics from the 20th century to crack open its secrets. Meanwhile, Sheldon Glashow has been a very outspoken critic of superstring theory almost from its very first year in the theoretical limelight. In a quotation from Martin Gardner's 1996 book *The Night is Large*, Glashow notes that 'I know that they are not going to say anything about the physical world that I know and love. Mainly, that's the reason that I don't like these theories... some of us at Harvard are still trying to follow the upward path, to go from experiment to theory, rather than pursuing the superstring vision, which requires the highest inaccessible dream-like energies to build a theory that deals with the down-to-earth world under our feet.'

Nobel Laureate Sheldon Glashow does not support superstring theory because it does not align with the world observed by experimental physicists who are the ultimate arbiters of which theory is correct.

Richard Feynman was also sceptical. In an interview with physicist Paul Davies published in the 1988 book *Superstrings: A Theory of Everything?* he reflects that: 'I have noticed when I was younger, that lots of old men in the field [of physics] couldn't understand new ideas very well.... I'm an old man now, and these are new ideas, and they look crazy to me, and they look like they're on the wrong track. Now I know that other old men have been very foolish in saying things like this... [but] I think all this superstring stuff is crazy and is in the wrong direction. I don't like that they're not calculating anything, I don't like that they don't check their ideas.'

Feynman went on to criticize the idea that superstrings work only in 10 dimensions, which means that an explanation must seemingly be artificially created to show why only 4 of the dimensions don't get compressed, instead of throwing the entire theory out as violating a very basic observational fact of our existence.

Although individual particles can be represented as vibrations of 1-dimensional strings within the compact 6-dimensional Calabi–Yau spaces, the unification of these fermion and boson particles requires supersymmetry as a key symmetry for converting the descriptions of fermions into bosons. That is why the discovery of supersymmetry is considered a key theoretical step in the entire superstring programme of unification. If evidence for supersymmetry is not found beyond the Standard Model, the entire 35+ year effort to refine superstring theory and fashion a Theory of Everything may have been for naught.

As for the topology of spacetime itself, the implications of string theory are less clear. The string-like character of all the fundamental particles does not become apparent until you reach a scale of 10^{-33} centimetres, or so the belief seems to be. Only then do strings stop looking like point particles. But this difference is enough to cure all string field theories of the infinities that plague point field theories such as QED and QCD. It is at this scale that it may not be correct to think about the strings as moving through what we normally consider to be continuous spacetime. Science popularizer and physicist Michael Green in his 1999 book *The Elegant Universe* explains further: 'In a theory of gravity, you can't really separate the structure of space and time from the particles which are associated with the force of gravity.... The notion of a string is inseparable from the space and time in which it is moving, and therefore if one has radically modified one's notion of the particle responsible for gravity, so that it is now string-like, one is also forced to abandon at some level the conventional notions of the structure of space and time...at these incredibly short scales associated with the Planck distance.... A lot of the present research is focused on trying to understand precisely how [this] works...'

By 1987, Steven Weinberg also viewed spacetime in the post-string world as a very different animal from what it used to be. In his book *Dreams of a Final Theory* he writes 'I think that in these theories space and time may not turn out to have overwhelming importance. Space and time coordinates are just four out of the many degrees of freedom that have to be put together to make a consistent theory, and it's only we human beings who give them that peculiar geometric significance which is so important to us.... What we are going to have is not so much a new view of space and time, but a de-emphasis of space and time. I don't think that everyone should work on superstring theory, and I don't think that everyone should work on phenomenology and low energy physics. I think people ought to do what they can... I think it's also worth trying to look ahead, jump over 17 orders of magnitude in energy and look up to the Planck scale where the final answer may lie.'

Key Points

- String theory proposes that all particles in the Standard Model are actually loops of 'string' which vibrate in 11-dimensions to give rise to the known properties of particles.

- There are five different string theories that operate in 10 dimensions, in which four of these are the usual 4-dimensional spacetime of relativity, but the remaining dimensions are small and form what are called Calabi–Yau geometries.

- The geometric properties of these 6-dimensional Calabi–Yau spaces determine the number, families and properties of the Standard Model particles.

- M-theory combines the five original superstring theories into a larger M-Theory that operates in an 11-dimensional spacetime, which is called the Bulk.

- The ensemble of possible superstring solutions is called the Landscape, and our universe is one among a vast number of these possibilities in an arena called the multiverse.

- Superstring theory is not a quantum theory of gravity because it depends on a pre-existing and background-dependent mathematical description that works only if gravity is not quantized and is very weak.

CHAPTER 13
Quantum Gravity

You will note from the discussions in the last few chapters that we have been obsessed with finding a GUT for the strong, weak and electromagnetic forces, but have ignored gravity altogether. This isn't an oversight. Gravity in the guise of 4-dimensional spacetime was considered to be pretty irrelevant compared to the strong, weak and electromagnetic forces, and the various Higgs fields that litter the landscape of spacetime. Spacetime served only as a convenient background providing a versitile coordinate framework for the benefit of keeping track of the other fields. Taking a clue from theories we have for the other three forces, it stands to reason that in order to create a Theory of Everything which includes gravity, we will have to create something like a quantum theory for the gravitational force and its field.

The search for a quantum theory of gravity has a long and complex history. During most of the last 200 years, only gravity and electromagnetism were known. Both of these forces followed the same 'inverse-square' law of diminution and seemed to have an infinite range of operation, which led many people to craft unified theories for them, including at one time or another Immanuel Kant (1724–1804), Michael Faraday (1791–1867) and James Clerk Maxwell (1831–1879). By 1786, Immanuel Kant even imagined that the universe was filled with forces so that each point in space contained attractive and repulsive agencies whose conflict produced all the phenomena of the observable world. From Kant's perspective, the forces of Nature such as gravity and magnetism, were merely manifestations of one another and could be converted, one into the other, given the proper physical conditions could be established. As it turned out, only electro-static and magnetic fields could be so easily interconverted, but it took another 34 years for someone to figure out how to perform this trick. That someone was André Ampère who discovered that two wires suspended side-by-side, in which currents flow in the same direction, repel one another just as the south poles of two magnets placed next to each other. Thus, magnetism could be produced from charged particles in motion (electric currents).

Michael Faraday, after many years of painstaking searching, finally demonstrated the existence of a new electric phenomenon in 1831: Induction. Each time the electrical current in a wire was switched on or off, or abruptly changed in strength, a weak current would begin to flow in the neighbouring wire. This led to a second discovery that a moving magnet could also induce such currents. In his 1852 treatise *Experimental Researches in Electricity*, Faraday described the process of magneto-electric

Immanuel Kant was an early proponent of the idea that matter is an epiphenomenon of nature and not itself fundamental.

induction in terms of 'lines of magnetic force'. He believed that these lines of force were 'strains in space' and that the laws of Nature were a consequence of the interactions of these strains.

Maxwell succeeded in translating Faraday's picturesque ideas of lines of force into a rigorous and mathematically precise theory. He was able to show that the electric and magnetic fields were part of a single mathematical entity he called the Electromagnetic Field. Instead of three forces in nature, gravity, electric and magnetic, there were only two: gravity and the new force of electro-magnetism. Maxwell had also discovered that whenever a disturbance was produced in an electric or magnetic field, this disturbance travelled through space in the form of an electromagnetic wave. That these electromagnetic waves were nothing more than ordinary light was later demonstrated by Heinrich Hertz (1857–1894). The invention by Guglielmo Marconi (1874–1937) of the 'wireless' telegraphy system quickly followed Hertz's discovery. Within a single human lifetime, these theoretical beginnings evolved into an avalanche of inventions including the radio, the television, and MTV.

At the time Maxwell and Hertz confirmed that light had wave-like properties, there appeared to be little confusion about how to think about light. Towards the end of the 17th century, Christiaan Huygens (1629–1695) had pioneered studies on the diffraction of light and had demonstrated quite convincingly that light was a wave phenomenon. At that time, all known undulatory phenomena were thought to require a medium to support their motion. It was quite natural for Huygens to propose that a substance, which they called the Ether, was responsible for the transport of light from one place to another analogous to water supporting a ripple on the surface of a pond. How else would the light from the sun travel through space unless there existed some transparent medium to carry it?

Natural philosophers by the 19th century recognized several forms of ether. The electric field required its own to function; magnetism a second; the gravitational field a third, and finally light needed a Luminiferous Ether to travel as a wave. Maxwell, in fact, thought of his electromagnetic equations as describing the tensile qualities of the ether as though it were a material medium like water or iron, which pervaded space, and even the spaces within every substance. Maxwell had succeeded not only in reducing electric and magnetic phenomena into aspects of the electromagnetic force, but unified three of the ethers into a single 'electromagnetic ether'. This wasn't the same Aether in vogue up until the time of Tycho because unlike the Aristotlean Aether, the Huygens-Maxwell ether could not have any resistive effect on planetary motion.

Life is simple when there are only two things to worry about. But, while the electromagnetic interaction was rapidly evolving from a classical, Maxwellian field theory into a relativistic quantum field theory, gravity during this same time seemed to languish. Much of the impetus for developing gravitation theory seemed to come more from the ranks of mathematicians enthralled by the abstract properties of Riemannian spaces. In contrast, much of the stimulus for advancing quantum electrodynamics, especially during its later years, came from physicists trying to understand a number of experimental results in atomic physics. With the exception of a handful of distinct experiments, general relativity remained, and still remains, a theory beyond the reach of modern experimental techniques. As Einstein once reflected, 'The theoretical scientist is compelled in an increasing degree to be guided by purely mathematical, formal consider-ations in his search for a theory, because the physical experience of the experimenter cannot lead him up to the regions of highest abstraction.'

There is also the question of just what is meant by unification. Is it sufficient for all fields to appear in the same equation, with symbols like A, B or C to denote them, or should fields appear as various internal flavours of a single super-field, D, such as D(A,B,C)? For that matter, if we were to follow a line inscribed on the fabric of spacetime, where do the points on that line cease to be space, and become the embedded quark or lepton? Since William Clifford (1845–1879) delivered his 1879 paper 'On the Space-Theory of Matter', and Einstein gave substance to this line of theoretical inquiry by his creation of general relativity, much effort has gone into creating a theory where matter is treated as a purely geometric phenomenon.

The seeds to this proposal can be found at least as far back as the 18th century. Roger Boscovich (1711–1787) and Immanuel Kant, for example, believed that matter was the result of the interactions between pure force fields acting upon each other in empty space; matter was merely an epiphenomenon tracing the interplays between far more fundamental events in the universe. Bernhard Riemann (1826–1866), in his 1854 paper 'Uber die Hypothesen welche der Geometrie zegrunde liegen' (On the hypotheses which underline geometry) was convinced that space possessed a structure (or topology) that was completely unlike the smooth, flat Euclidean geometry of Newtonian physics. Some years later, in 1872, the mathematician Felix Klein (1849–1925) began his Erlangen Program from which he, too, hoped to

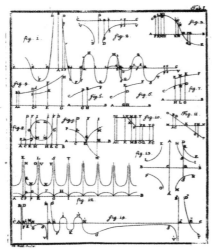

Image from Roger Boscovich's Theoria Philosophiae.

show that all physical laws and quantities could be described by purely geometric concepts.

That there was a peculiar, ephemeral aspect to matter was already appreciated both theoretically and experimentally as early as 1901, when it was discovered that the electron behaved as though *all* of its mass was due to its electromagnetic field alone. Instead of a massive, tiny sphere embedded in its own electrostatic field, the experiments demonstrated that the way the electron behaved upon acceleration was consistent with it being a purely electromagnetic phenomenon, with no massive tiny sphere left over. Modern experimental results also show that the proton consists of three quarks, but that the sum of the rest masses for the quarks of 20 MeV is 46 times less than the mass of a single proton. How can a body weigh more than the sum of its parts? The answer is that the balance of the mass is in the form of the energy latent in the gluon field surrounding each quark – a totally non-physical entity but a simple manifestation that confirms Einstein's equation $E = mc^2$. Energy and mass are just different sides of the same conceptual coin, only their total value, regardless of apportionment, is important.

Now, as you might well imagine, divesting matter of its mass and considering electrons and protons as persistent knots in some field has few practical applications, since physicists find it useful to think of even the smallest particles as miniature spheres embedded in an otherwise implacable empty space. Yet, the idea that matter is just another form of empty space, much as Einstein showed that mass is just another form of energy, has several profound implications which reach down to the very bedrock of physical reality. Towards the latter years of his life, Einstein was so certain of the correctness of the idea that matter is not a fundamental ingredient to nature, that he insisted: 'The material particle has no place as a fundamental concept in field theory. Even Maxwell's electrodynamics are not complete for this reason.' Gravity as a field theory must also deny a preferred status to matter. Another modern perspective on this subject is voiced by John Wheeler in the 1973 book *Gravitation*: 'What else is there out of which to build a particle except geometry (spacetime) itself?'

JOHN WHEELER

John Archibald Wheeler (1911–2008) received his Doctorate at Johns Hopkins University in 1933 and became a professor of physics at Princeton University in 1938, where he worked on models of the atomic nucleus and S-matrix theory until he was brought in to the Manhattan Project in 1942. He was instrumental in developing the hydrogen bomb in 1950. Among his students was Richard Feynman. Beginning in the 1950s, Wheeler developed the post-general relativity ideas of geometrodynamics, and was the key developer of many concepts in classical and quantum gravity theory, including coining the term 'black hole' and 'worm hole' along with the foamy conception of spacetime at the Planck scale. One of his most dramatic ideas was the participatory universe in which observers create the physical universe by their very act of observation.

John Wheeler

The theoretical program of the geometrization of matter, which began in earnest with Hermann Weyl (1885–1955) in 1924, was subsequently developed by John Wheeler (1911–2008) and Charles Misner (1932–) between 1955 and 1957. Spacetime at a scale of 10^{-33} cm was imagined, not as a smooth flat sheet of paper, but as one in which loops and bridges between one region and another appeared and vanished, and the geometry of space was highly curved and warped at this scale.

To see how this works, physicists work with the Planck units of scale, which are generally interpreted to indicate the scale at which the quantum effects of spacetime become significant. They are defined by using the Newtonian constant of gravity (G), the speed of light (c) and Planck's constant (h), in the appropriate combinations to create the appropriate physical units as follows:

$$Lp = \sqrt{\frac{hG}{2\pi c^3}} \qquad mp = \sqrt{\frac{hc}{2\pi c}} \qquad tp = \sqrt{\frac{hG}{2\pi c^5}} \qquad Tp = \sqrt{\frac{hc^5}{2\pi Gk^2}}$$

When the appropriate values are used for these constants, you get:

Planck length	$L_p = 1.6 \times 10^{-33}$ cm
Planck mass	$m_p = 2.2 \times 10^{-5}$ grammes
Planck time	$t_p = 5.4 \times 10^{-44}$ seconds
Planck temperature	$T_p = 1.4 \times 10^{32}$ kelvins
Planck energy	$E_p = m_p c^2 = 2.0 \times 10^{16}$ ergs or 1.3×10^{19} GeV

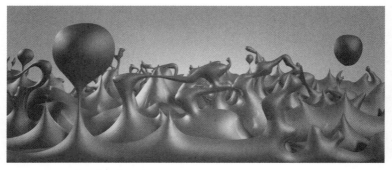

Spacetime at the planck scale.

Wheeler took this idea one step further and described spacetime in which the geometry of every possible 3-dimensional space (called by physicists 3-space) is assigned a probability, and Heisenberg's Uncertainty Principle ensures that our 3-space is just an average of innumerable quantum topologies at any one instant. Each 3-space geometry contains a set of quantum fields – one for every non-gravitational field known – which act together to distort the flat geometry of space into some shape consistent with Einstein's equations. The geometric details of these 3-spaces are subject to fluctuations at the Planck scale, and our universe apparently shifts between one of these geometric states and other similar ones, along a path of maximum probability. Each of these geometric states was thought of as similar to the wave function of an electron so that just as the quantum states of an electron were elements of a Hilbert space of all possible wave function states, the quantum states of our 3-dimensional space were elements of a so-called Wheeler Superspace of possibilities. Just as the evolution of an electron could be described as a path through Hilbert space connecting individual quantum states according to Schrödinger's Equation, the evolution of our 3-dimensional space could be traced through superspace as a path linking the individual geometric quantum states. The equation that described this evolution was developed by Wheeler and Bryce DeWitt (1923–2004) and is called the Wheeler–DeWitt equation.

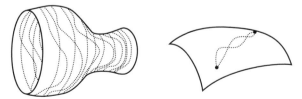

Spacetime represents one 3-dimensional geometry (g1) changing into another geometric state (g2). Superspace contains all possible geometric states for 3-dimensional space dictated by the locations of the internal matter fields, and spacetime is created by connecting these individual states using the Wheeler–DeWitt equation.

Relativists weren't the only ones thinking about what to do with gravity. Other physicists steeped in the language of quantum mechanics were taking a very different and perhaps more pragmatic approach. By the 1930s a quantum theory for the electron had been fashioned by Dirac. More importantly perhaps is that, with the discovery of the strong and weak nuclear forces, nature no longer looked quite as simple as it had with only gravity and electromagnetism to worry about. This pretty well spelled the end to dealing with electromagnetism and gravity based on a classical approach to field theory and unification. In the card game of theoretical physics, the ante had now been raised so that only quantum field theory could serve as an acceptable starting point for such pursuits. With a quantum theory for gravity nowhere in sight, attention turned away from gravity unification and towards more profitable areas of investigation.

Still, some physicists struggled onwards as best they could with only a sketchy guide of the landscape of quantum gravity to guide them. One of those was Léon Rosenfeld (1904–1974) who, in 1930, attempted to apply the fledgling techniques of quantum field theory to the gravitational force by calculating the gravitational self-energy of the photon. He was the first to identify many of the technical problems that such quantum gravity calculations would have to overcome. Not the least of these was an even more terrible plague of infinities than anyone had ever seen in quantum electrodynamics. DeWitt, in his 1967 paper 'Quantum Theory of Gravity', pointed out that 'Rosenfeld's result [was] a forecast that quantum gravidynamics was destined, from the very beginning, to be inextricably linked with the difficult issues lying at the theoretical foundation of particle physics.'

As part of his PhD thesis at Harvard University, DeWitt re-performed Rosenfeld's calculations 20 years later, spurred on by the significant advancements that had been made in QED since the time Rosenfeld attempted to tackle the same calculations. Even though the QED infinities had been shackled using the renormalization technique, DeWitt's results only uncovered still more problems with naively applying QED techniques to gravity. Renormalization would have to be applied not just twice as in the case of QED (to redefine the mass and charge of the electron), but over and over again an infinite number of times just to keep the gravitational calculations from blowing up. These early efforts by Rosenfeld and DeWitt led to a vastly improved understanding of what a quantum gravity theory would have to consist of, and a first reconnaissance of this theoretical landscape, but they also yielded not a shred of evidence for what a true theory should look like that could consistently explain gravity as a quantum field. According to physicist Joshua Goldberg (1925–) in his 1984 book *Gravitation, Geometry and Relativistic Physics*, 'The quantization of the Einstein theory ... raises fundamental questions about the meaning of geometry, of spacetime, of manifolds, and of the relationship of gravitation to the rest of physics.'

Bryce DeWitt along with John Wheeler developed the key ideas behind the quantum theory of spacetime by using ideas from ordinary quantum mechanics such as the state of a system.

During this century, the search for a theory of quantum gravity has followed one or the other of two courses of investigation. In the first approach, called Covariant Quantum Gravity, the problem of what to use as a background upon which to hang the gravitational field was solved by breaking spacetime into two parts. One part would represent a flat spacetime, and a second part would act as the dynamic element of the gravitational field. In other words, the second component could be treated like an ordinary field that only weakly disturbs the background geometry of spacetime. This means that after all the mathematical machinery of quantum gravity has run its course, the background field used in Covariant Quantum Gravity must utterly vanish as a completely undetectable scaffolding to the theory – an even more thorough disappearing act than performed by virtual particles in QED. This is the approach taken by superstring theory.

The second approach, called Canonical Quantum Gravity theory, has a very different parentage. Following the established mathematical techniques developed by Schrödinger and Dirac, Einstein's equation for gravity gets rewritten, and it is in this new form that it can be analyzed following the conventional mathematical techniques in ordinary quantum mechanics. This forces quantum gravity theory to come out looking like a respectable quantum theory, but one in which the dynamics of 3-space play a key role, and the new formulation describes how a system, the complete 3-space geometry of the universe, changes from state to state. Whereas quantum mechanics depends on considering the histories of particles as they move through spacetime, quantum gravity must, in addition, consider all possible geometries of space as they unfold in time. This is the approach taken by Wheeler and DeWitt.

By the 1960s, several new attacks were launched on the theoretical underpinnings of the quantum gravity programme by Richard Arnowitt (1928–2014), Stanley Deser (1931–), and Charles Misner (called the ADM theory). In ordinary quantum field theory, each field is considered to be superimposed upon 4-dimensional spacetime and is characterized by a particular quantum configuration. This configuration is indexed by a unique set of so-called quantum numbers that represent the associated particle's spin, energy and angular momentum. This configuration changes in a well-defined way as the system evolves subject to external influences such as absorbing a photon. Each state would be assigned a probability, and the evolution of the system would be dictated by a fundamental 'probability wave' equation. The electron is described as a wave whose shape depends on its specific energy, angular momentum and spin. All of the possible quantum configurations of the electron are elements of a new kind of space called Hilbert Space. This is an infinite-dimensional space, one dimension for each of the quantum states of the electron. As the electron's wave function changes, the electron moves from one configuration to another in Hilbert space.

So how should we carry over these basic ideas of quantum mechanics into quantum gravity? What will take the place of a quantum state, a wave function, or even Hilbert space? And what properties of the gravitational field and spacetime would take the place of the discrete quantum numbers that were previously used to index all of the infinitude of possible quantum states of a particle? The answers provided by Canonical Quantum Theory are both surprising and profoundly unsettling if true.

The equation that represents the evolution of the particle system is now an equation that describes how an entire 3-dimensional geometry for the universe changes from moment to moment. Time, in this sense, is just a new variable that indexes each of the 3-geometries of space. Just as quantum mechanics required Hilbert Space to keep track of the possible configurations of a quantum system in space, in the ADM theory, quantum gravity will also need its own analogue to Hilbert Space in which every single possible state of the universe's 3-dimensional geometry is represented. This arena for quantum gravity is called Superspace. As John Wheeler puts it in the 1973 book *Gravitation*: 'Superspace is the arena of geometrodynamics. The dynamics of Einstein's curved space geometry runs its course in Superspace as the dynamics of a particle unfolds in spacetime... The track of development of 3-geometry with time [is] expressed as a sharp, thin "leaf of history" that slices through Superspace. The quantum principle replaces this deterministic account with a fuzzed-out leaf of history of finite thickness. In consequence, quantum fluctuations take place in the geometry of space that dominate the scene at distances of about the Planck length.'

This analysis also leads to some remarkable inferences. For example, the Heisenberg Uncertainty Principle denies the existence of a fixed classical spacetime. Spacetime only has a meaning in an average sense and at scales much larger than the Planck scale of 10^{-33} cm. Space and time themselves, are not primary but are secondary ideas in nature and serve merely as approximations to some perhaps more underlying reality. In 1968, Wheeler noted in his article Superspace and the Nature of Quantum Geometrodynamics that, '...they have neither meaning nor application under circumstances where geometrodynamic effects become important... There is no spacetime, there is no time, there is no before, there is no after. The question of what happens "next" is without meaning.' As you approach the Planck scale at 10^{-33} centimetres, spacetime becomes awfully contorted, but at the same time you never seem to run out of spacetime coordinates themselves. Space develops wormhole handles, but the surfaces of these handles consist of their own sub-Planckian spacetimes patched together into an increasingly distorted spacetime. By the late 1960s, the Canonical Quantum Gravity or Superspace Quantization approach was bogged down in an ever-growing mathematical formalism in which the equations defining the quantum states of spacetime became too difficult to solve in general. Covariant Quantum Gravity, on the other hand, with its reliance on concepts refined in the study of particle physics received several infusions of fresh blood during the '60s.

Richard Feynman and Bryce DeWitt worked on systematically applying the techniques of QED – the so-called perturbation series approach – to gravity. This simply meant that they started with a very simple Feynman diagram for, say, a photon emitting a single graviton, then two gravitons, then three, etc. The contributions at each stage were added up to give the total strength of the process that involved a single photon interacting with a cloud of virtual gravitons. From this mathematical approach, they were able to prove that whatever quantum gravity theory may look like, it was at least a unitary theory, meaning that for every new term added to the series, probabilities would continue to sum to exactly 1. This was at least true for weak gravitational fields. For strong fields it was quickly discovered that gravity continued to be non-renormalizable. A complete calculation of some process in this limit would spawn more and more quantities that needed to be renormalized. At least in QED the renormalization approach only led to the redefinition of two quantities, the electron mass and charge.

The search for a renormalizable quantum theory of gravity led to many new attempts at creating a new family of renormalizable quantum gravity theories by adding new terms to Einstein's original equation for gravity. Einstein had already added the cosmological constant term to his equation for gravity in order to stabilize the universe against expansion or collapse. Similar terms were added to stabilize gravitational calculations against the scourge of infinity. This work was, however, overtaken by developments in the search for GUTs and the unification of GUTs with gravity using supersymmetry theory, described in the previous chapter.

Apart from the ambiguous experimental evidence that gravitational radiation exists but its quantal effects are unmeasurable, what compelling theoretical evidence do we really have that gravity should be expressible as a field theory similar to QED, QCD and Electroweak theory, which form the basis of the Standard Model? If you suppose that gravity is mediated by a quantum particle called the graviton, the graviton must satisfy some very basic properties. We have already seen how the graviton has an upper limit to its possible rest mass, which is 100 trillion times lower than the photon itself. In addition to being massless, physicist Steven Weinberg proved that the graviton must have a spin assignment of 2 units so that the equation defining it looks like Einstein's equation for the gravitational field represented by a two-index field, $g_{\mu\nu}$. Weinberg also went on to prove an important theorem in 1964, which showed that spin-2 particles have to interact with all other particles and fields with a universal strength.

So, even without a single clue to what such a theory ought to consist of, we can already specify with some certainty what the properties of the quantum of gravity ought to look like. The graviton has all the right properties as a theoretical particle to be the carrier of gravity; now all we have to do is create a field theory to go along with the particle. This takes us smack into two hard questions: can such a theory of gravitons be described by the same class of theories that have proven so effective in the Standard Model? Also, can it be shown that graviton quantum field theory is free of infinite answers, just as QED and QCD are now known to be, thanks to the renormalization technique?

In 1956, Ryoyu Utiyama (1916–1990) at the University of Osaka showed that, just as in the case of electromagnetism and Yang–Mills theory, gravity can also be expressed as a similar kind of field. This was a significant new finding for gravity research since it confirmed that gravity was of the same class of theory as ones that were already proving to be important as the foundation for understanding the other three forces in the Standard Model. When you think about it, this is a rather astonishing result because gravity certainly seems to look different than the other forces. By 1963, Bryce DeWitt went on to suggest that Yang–Mills theory might be unified with gravity by adding a 5th, very-small new dimension to spacetime. The problem with this approach turned out to be that there were no good reasons why the new dimensions that had to be tacked on to 4-dimensional spacetime should be tiny. Still, apart from this technical difficulty, it looked like a unification theory of all forces was close at hand. Had a convincing quantum theory of gravity at last been spotted? Not yet. A collection of ideas is not the same as a rigorous mathematical framework from which real quantitative predictions can be launched.

Yang–Mills theory applied to gravity still produces infinities that cannot be tamed using the by then standard renormalization techniques. A variant of this approach was invented by Gerard 't Hooft and Martinus Veltman (1931–), called dimensional regularization, in which the calculations were first carried out in a spacetime with greater than 4 dimensions, then converted back to a 4-dimensional answer. For calculations involving pure gravitons interacting only with themselves, this technique provided finite answers. The only problem was that it required that more and more terms be added to the defining equations of the theory as each new field was included, and this made the whole theory look ad hoc and inelegant.

By the early 1970s, the subject of quantum gravity had reached something of a crossroad. The basic problems involved in creating such a theory along the lines of other familiar quantum field theories were pretty well understood, but there were no answers as to how to circumvent them mathematically. Most significantly, the infinity problem was a glaring indicator that something was lacking in the approaches being taken. Beyond this, there was still no genuine unanimity over which approach to take as the starting point for actually creating a quantum gravity theory. For gravity, there were many features in general relativity that could be quantized. General relativity posits spacetime as a collection of points or events that form a 4-dimensional manifold whose shape is determined by Einstein's equation for gravity. You could imagine trying to quantize the 'points', the 'manifold' or the gravitational 'field'. Then again, could it be that spacetime was itself too small an arena? The extension into 5 dimensions had been tried and discarded because it didn't lead to a sensible theory. How about the relativistic equation of gravity proposed in general relativity? How much of it should be subjected to the quantization approach?

$$R_{\mu\nu} - \frac{1}{2} R\, g_{\mu\nu} + \Lambda\, g_{\mu\nu} = \frac{8\pi G}{c^4}\, T_{\mu\nu}$$

The right side of the equation contained an accounting of the mass-energy in spacetime, but these were known to be quantum fields. If these are quantizable, then shouldn't all of those geometric quantities on the left side of the equals sign also be quantizable? But which ones? Should R, $R_{\mu\nu}$ and $g_{\mu\nu}$ be quantized together, or just one of them? How about the other geometric fields that are possible in the full Riemannian theory of geometry such as $R_{\alpha\mu\beta\nu}$ or combinations of all of these quantities that represent different aspects of curvature and the gravitational field? Evidently, not even Einstein's own equation for gravity was comprehensive enough when attempts were made to recast it in the language of quantum mechanics.

Key Points

- Quantum gravity is a quantum theory of gravity and spacetime which describes the physical universe at the Planck scale, where many of the techniques of quantum mechanics may be applied to describing spacetime.

- A number of approaches to developing a quantum theory of gravity have been attempted over the last 100 years but all have failed to create a completely unified and accurate description of spacetime at the Planck scale.

- These attempts have led to a growing vocabulary of terms that describe the technical properties of spacetime as a quantum phenomenon.

- Attempts at developing a quantum theory of gravity along the lines of QED or QCD have led to powerful insights into the mathematical problem of treating spacetime as a quantum field.

- It is expected that at the Planck scale, the geometry of 4-dimensional spacetime as a definite physical object will transition into an object that manifests the indeterminacy of a quantum object in which its geometry changes randomly and across complex topologies.

CHAPTER 14

Evidence for Quantum Gravity

Although theoreticians working in this difficult area of physics are quite bullish about its chances for success – at least eventually – the critics of the quantum gravity programme have never been this certain about the inevitability of finding such a Mother of All Theories. Richard Feynman in his *Feynman Lectures on Gravitation* stated in the 1980s: 'The extreme weakness of quantum gravitational effects now poses some philosophical problems: maybe nature is trying to tell us something new here, maybe we should not try to quantize gravity...it is still possible that quantum theory does not absolutely guarantee that gravity has to be quantized.'

Physicists who are suspicious of the quantization of gravity programme often argue that quantum mechanics works well when you have test particles like electrons that are vastly smaller than the scale of the quantum phenomena. But when you reach the Planck scale at 10^{-33} centimetres, you run out of test particles, measuring sticks, and clocks that are smaller than the phenomena for which you are attempting to formulate quantum laws. Without a means to measure distances and time intervals you cannot describe the motion of particles through space.

A quantum black hole with a mass equal to $M_{pv} = 2.2 \times 10^{-5}$ grammes will have a radius of $L_p = 1.6 \times 10^{-33}$ cm, and will decay by the emission of Hawking Radiation in a time equal to about $t_p = 5.4 \times 10^{-44}$ seconds. In the figure on the previous page, what this

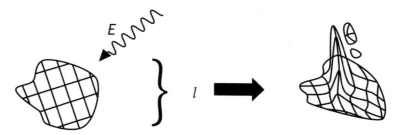

Example of measurement at the Planck scale, where the incoming photon perturbs spacetime so that its prior geometry cannot be determined due to the Heisenberg Uncertainty Principle.

means is that, if you attempted to probe the structure of spacetime at a scale of the Planck length L_p, you would need to use a photon with a wavelength of λ, which carries an energy equal to the Planck energy, E_p.

Such a photon carries so much mass that it immediately becomes a quantum black hole with a radius of L_p, and evaporates into a hailstorm of energetic particles within a time equal to t_p. That means that any information the photon obtained from spacetime structure at this scale is immediately lost in a completely randomized burst of particles for which their position and momenta are also scrambled randomly. What this implies is that, by the time you arrive at the Planck scale, you have reached an epistemological horizon to nature that is beyond measurement and mathematical formulation in terms of known quantum principles, or even techniques of measurement and observation.

Apart from the epistemological problem and the many failed attempts at creating a quantum theory of gravity, is there any shred of physical evidence that such a programme might even be needed? The evidence is by no means compelling. The only other long-range force we know about is electromagnetism. Maxwell's theory showed that the electromagnetic field can support self-sustaining waves – electromagnetic radiation. The work by Planck on the blackbody spectrum and by Einstein on the photoelectric effect convincingly showed that this field is quantized in terms of photons. Are either of these experimentally proven conditions for the electromagnetic field – namely, its wavelike behaviour and its quantization – also true for the gravitational field?

HULSE-TAYLOR BINARY PULSAR

The good news is that gravitational radiation emission has been confirmed by careful monitoring of certain objects in deep space called pulsars. In 1974, John Taylor (1941–) and Russel Hulse (1950–) at the University of Massachusetts had been studying a binary star system consisting of two neutron stars which were themselves detected as pulsars, and designated by the cryptic name PSR 1913+16 – or simply the Hulse–Taylor binary. Careful monitoring of this binary star system over a period of 10 years provided spectacularly accurate measurements of their orbital elements and their changes in time. For instance, the neutron stars orbit each other in a year whose length is 27906.9807807 seconds (approximately 7¾ hours).

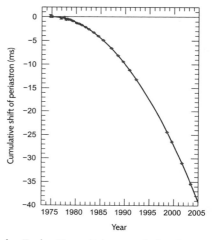

General relativity predicts that such systems should emit gravitational radiation due to the tremendous accelerations these stellar-massed bodies are experiencing. In fact, by 1981 their orbital period was found to be decreasing; their orbital 'year' was getting shorter, by 2.425 trillionths of a second per second. Does this match the prediction of general relativity?

Hulse-Taylor Binary Pulsar graph showing how the energy of the system is decreasing in time as they emit gravitational radiation.

Apparently so. The ratio of the general relativistic prediction to the actual decrease in the orbital period is 1.0023 +/- 0.0047 where 1.0000 is an exact verification. No other known effects appear to be capable of mimicking the apparent loss of energy from the system due to gravitational radiation, and a good many effects have been considered. The fact of the matter seems to be that, like a leaky bag, PSR 1913+16 loses as much energy in 1 hour in the form of gravitational radiation as the sun loses to normal light radiation in about 2 minutes. This spectacular result, and indirect proof for the existence of gravitational radiation, earned Hulse and Taylor the 1993 Nobel Prize in Physics.

GRAVITY WAVES

Physicists had searched for signs of gravity waves since the 1960s when Joseph Weber (1919–2000) at the University of Maryland built a simple gravity wave detector. It consisted of cylinders of aluminium with strain gauges capable of registering displacements as small as 10^{-16} metres as a gravity wave travelled by. Despite decades of searching, no gravity waves were ever detected by this method. However, Weber is widely regarded as the Father of Gravity Wave Detection because he was the first to make the attempt in a serious way, and this paved the way for others to follow his lead in this difficult area of research.

JOSEPH WEBER

Joseph Weber (1919–2000) graduated from the US Naval Academy in 1940 and served aboard many ships during the Second World War, reaching the rank of Lieutenant Commander. In 1948 he resigned his commission and joined the engineering faculty at the University of Maryland, completing his engineering PhD at the Catholic University of America in 1951. He was nominated for a Nobel Prize in 1962 for his ground-breaking work on masers but the award was given to others in 1964. Meanwhile, his other interest in general relativity led him to the design of the first gravity wave detectors in the late 1960s, which, unfortunately, never detected such waves although they set stringent limits to their intensity at the wavelengths he measured.

Joseph Weber working on a gravitational wave detector.

Eventually, new technologies appeared with far-better sensitivities than Weber bars, at least to low-frequency gravity wave signals from colliding neutron stars and black holes. Kip Thorne (1940–) at the California Institute of Technology led the way in developing a thorough theoretical understanding of how gravity waves behaved, what kinds of astrophysical systems would produce them, and an evaluation of various technologies for detecting them. Interferometers were studied and several detectors were built between 1970 and 2000. Rather than measuring the physical strain and displacement in a solid mass, interferometers used lasers in a fixed geometry of the Michaelson design. If the paths travelled by the two laser beams was slightly different, this would cause movement of interference fringes, which could be detected by a photoelectric device.

In 1988, the Laser Interferometer Gravity-Wave Observatory (LIGO) Project received start-up funding to design a pair of large instruments in Hanford, Washington and Livingston, Louisiana. The actual construction started in 1994. At a cost of $1.1 billion, the twin instruments consist of a pair of mirrors on two 90-degree cross arms separated by 4 kilometres from a central metrology and

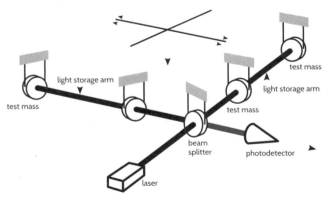

The LIGO gravity wave interferometer measures changes in the geometry of space over one million times smaller than the diameter of an atom, and by 2020 has detected over 40 gravity wave events caused by colliding black holes and neutron stars..

beam-combiner laboratory. A 20-watt laser system is amplified so that the power travelling through the mirror system is over 700 kilowatts. The beams are combined to produce interference fringes at the central station and nulled so that when there is no distortion to the beam geometry, a minimal light signal is detected. Any disturbance to the geometry of the mirror system by a passing gravity wave causes fringes to shift slightly and the original interference condition to be disrupted, causing more photons from the laser to be detected by the photometer system. The sensitivity of the system is such that a displacement of the mirrors in the interferometer by 10^{-18} metres can be detected.

Initial observations began in 2002 and continued with several instrument upgrades to sensitivity, culminating in the first detection of a gravity wave pulse, GW150914, on 14 September 2015. The detection was announced in a major press release and publication in the journal *Nature* on 11 February 2016, in the paper Einstein's Gravitational Waves Found at Last. Following this initial discovery, many more events were detected by 2019, with

The control room of the Laser Interferometer Gravity-Wave Observatory at Livingston.

a catalogue of events that now numbers over 45, including events related to neutron star-neutron star collisions and mergers, along with neutron star-black hole and black hole-black hole mergers.

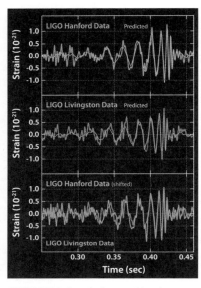

GW150914 signal observed by the twin LIGO observatories at Livingston, Louisiana, and Hanford, Washington.

Although gravity waves have now been verified and brought into the mainstream of gravity research, this places us at an equivalent time of 1890. Electromagnetic waves had been detected in 1887 by Heinrich Hertz, and then the quantization of this radiation was detected a few decades later through the efforts of Max Plank and Albert Einstein. The bad news is that no one has ever seen a graviton, nor any natural phenomena requiring such a particle – at least, not yet. There may, however, be indirect ways to detect them.

GAMMA RAYS AND QUANTIZED SPACETIME

If gravity is a quantized field of nature related to, and identical to, spacetime, then gravitons represent spacetime being itself quantized, presumably at the Planck scale of distances near 10^{-33} cm. When photons of ordinary light travel through such a quantized spacetime, their cumulative paths should differ due to their random walk from one portion of quantized spacetime to another along their journey. Moreover, the wavelength of light (e.g. its energy) should set a scale at which the photon is exposed to the Planck scale irregularities. A low-energy, long-wavelength photon would 'see' a much smoother spacetime geometry than a high-energy, short-wavelength photon. What this means is that for a collection of photons of different energy travelling long distances through quantized spacetime, their arrival times should be longer for the short-wavelength photons and shorter for the long-wavelength photons. This implies

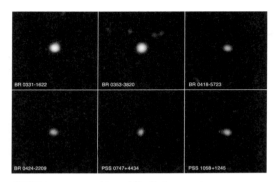

GHRB images used to set limits to quantum spacetime foam.

that the speed of light is dependent on the wavelength of light, which violates the basic Lorentz Invariance principle of special and general relativity which states photons always travel at the speed of light, no matter what their wavelength. But how do we create an experiment in which this energy dispersion can show up? The answer is that it may already have been conducted.

On 16 September 2008, NASA's Gamma-ray Large Area Space Telescope (GLAST, now renamed Fermi), recorded the most powerful gamma-ray bursts (GRBs) observed by 2008. The burst (known as GRB 080916C) took place at a distance of 12.2 billion light years, and indeed the highest-energy photons arrived about 16 seconds later than the lower-energy photons. However, so little is known about the emission mechanism of GRBs that such time delays may say more about the origin process than any possible quantization of spacetime along the gamma-rays journey. By 2015, an investigation of six additional GRBs led Eric Perlman at the Florida Institute of Technology and his colleagues to determine that the data was incompatible with a model in which the gamma rays diffuse through spacetime the way light diffuses through fog. The images of the GRBs should be far larger and fuzzier than they are found to be, and this also set a limit to the spacetime structure, which was smoother and less 'foamy' than many quantum gravity models have predicted. Further studies of GRBs may eventually let astronomers disentangle the issue of time delays in the origin mechanism and Lorentz violation in the quantization effect. Meanwhile, other ingenious experiments have been proposed that could also be sensitive to space quantization.

In one set of experiments, the fluctuations of quantum spacetime cause the noise in a laser interferometer to be different than for a

Gamma-Ray Large Area Telescope.

static background space. Since 2014, Craig Hogan at Fermilab in Illinois has been operating such a gravity Holometer. The instrument uses laser beams to detect the quantum noise in the system, and can identify departures from the predicted noise due to the addition of a quantized spacetime effect. By 2015, it was claimed that the Holometer had ruled out quantum spacetime fluctuations, but both the null result and the operating principles of the instrument itself are considered controversial.

Another set of experiments were proposed in 2017 by Sougato Bose at University College London and Chiara Marletto at Oxford University, which explores quantum gravity indirectly. In Chapters 4 and 5, we saw that the key feature of quantum mechanics is the property of super-position and entanglement. If gravity is a quan-tum system, then it must support entangled gravitational states. The experiment consists of preparing two

Graph of experiment sensitivity to fluctuations in spacetime.

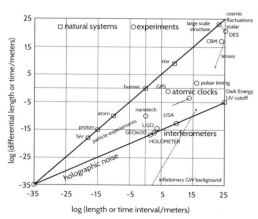

microscopic collections of matter called microdiamonds, within whose interstices are embedded conventional atoms of nitrogen. The microdiamonds are then dropped in a gravitational field so that their mutual gravitational fields can place them in an entangled state. A careful study of the state of the nitrogen atoms will determine whether the diamonds were in an entangled state or not. If so, gravity must behave as a quantum field even though we cannot detect the individual quanta of the field directly.

Finally, another indirect 'proof' that spacetime is quantized comes from the study of black holes, information theory and thermodynamics. Investigations started by Stephen Hawking (1942–2018) in 1975 have led to the idea that the surface area of a black hole is proportional to its entropy, which is a measure of its information content. Through what is called the Holographic Principle, one 'bit' of information corresponds to an area on the surface of a black hole's event horizon that is $A_p = L_p^2$ or 10^{-66} cm^2. The event horizon is a 2-dimensional surface which acts like a hologram and contains all of the information represented by the matter and internal volume of the black hole. Since the event horizon has a finite surface area, it contains a finite amount of information in A_p units, and so the volume must also consist of a finite number of quantum-scale volumes. If space were infinitely divisible, the surface area of the event horizon would also have to be infinite, which it is not.

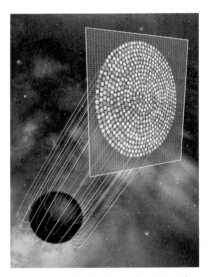

According to the Holographic Principle, all of the information that defines a 3-dimensional object is encoded in a 2-dimensional surface. This relates to the information within a black hole and its relationship to the event horizon.

Key Points

- A quantum theory of gravity would propose that gravity behaves much like other fields and can generate its own radiation from accelerating masses called gravitational radiation.

- Gravity waves have been indirectly detected from the Hulse-Taylor binary pulsar, and directly detected in 2015 at the LIGO observatory.

- The proof that spacetime is actually quantized may have been found from studies of gamma ray bursts, in which the higher-energy photons arrived significantly later than the lower-energy photons.

- A number of other ingenious experiments to detect signs of a quantized spacetime have been attempted, but as yet there is no convincing evidence that spacetime is quantized.

CHAPTER 15
Loop Quantum Gravity

As we saw in Chapters 8 and 9, physicists were anxious to develop an all-encompassing theory that combined all of what we know about the Standard Model and gravity into a comprehensive and self-consistent mathematical theory. The theory had to be able to treat each of the four fundamental forces in a unified description of how they operated. Moreover, it had to specify the kinds of particles responsible for mediating these forces as well as the particles they acted upon. The theory had to do so in such a way that all possible interactions and properties of the particles could be calculated from the theory, giving finite predictions that matched one for one the known properties of particles and the outcomes of their interactions.

The discovery of the quantization of light and matter leading to the relativistic theory of the quantum field led to the creation of the first of these intermediate theories: QED. This was followed rapidly by the quantum theories of the strong and weak interactions, and the first synthesis of two of these theories unto a unified electro-weak theory in the late 1960s. These theories were all found to be aspects of specific symmetries in nature classified as the simple unitary group $U(1)$, and the special unitary groups $SU(3)$ and $SU(2)$ for the strong and weak interactions. Together, these quantum field descriptions for the three forces, together with the small number of elementary particles, led to the Standard Model with its 12 spin-1 bosons that mediate the three forces, and the 12 spin-½ fermions (electrons, neutrinos, quarks) that represent all known forms of matter. A thirteenth particle, the Higgs boson, has spin-0 and provides the symmetry-breaking mechanism that ultimately confers the property of mass onto all of the other particles, and causes the electro-weak force to separate into two separate forces at low energy.

There are, however, problems with the Standard Model. It does not represent a truly unified mathematical picture of how the forces operate. Although all three mathematical descriptions are technically classified as what were called renormalizable non-Abelian gauge field theories, the electro-weak part of the model is represented by one kind of theory while the strong force is represented by QCD. Attempts at unifying these descriptions into one 'grand unified theory' have met with a variety of technical and observational difficulties, although much has been learned from the many attempts at unification within the Standard Model.

When attempts are made to bring gravity into the Standard Model by quantizing it in terms of spin-2 gravitons, the mathematics leads to severe non-renormalizable calculations and the infinities even for simple two-particle graviton-to-graviton interactions. This is because gravitons are generated by any source of matter or energy via $E = mc^2$. Calculations quickly spawn numerous additional gravitons that also have to be dealt with, and which generally produce infinite non-converging answers for interaction rates and momentum changes. Also, since the 1940s, calculations in quantum field theory have relied on the weakness of the forces so that any calculation can be viewed from a perturbation approach where you sum together successive interactions involving 1, 2, 3… quanta, and these contribute to the total sum with declining numerical importance. With QED, each new photon added contributes $(1/137)^2$ to the probability and so this perturbation approach converges to a finite answer very rapidly. For QCD the factor is $(0.6)^2$, so many more terms, literally millions, have to be summed to converge. For attempts at gravity quantization, this perturbation scheme fails completely because each term itself represents the generation of an infinite number of gravitons! There is no known way to use the standard perturbation and renormalization approach to fashion a quantum theory of gravity, mainly because there is a deep incompatibility between quantum mechanics and general relativity.

Quantum theory and the Standard Model are expressed in terms of fields operating within a flat, 4-dimensional spacetime identical to the one used in special relativity. Even superstring theory operates within a flat spacetime. Though graviton-like particles are included in superstring theory, they are treated as weakly interacting with the pre-existing background spacetime, which remains flat. In other words, the most theoretically successful field theories do not apply to conditions where gravity is strong and spacetime is warped. The hope is that string theory will eventually be found to work in such a way that the spin-2 strings can be made to generate their own 4-dimensional spacetime so that the pre-existing one can be eliminated as scratch paper. String theory and quantum mechanics are, therefore, called background-dependent theories.

Meanwhile, general relativity, which has demonstrated its accuracy as a classical field theory for gravity, is not only not a quantum field but is a purely relativistic one that does not require a pre-existing spacetime framework. Because of the relativity principle, it is a background-independent theory that literally creates the properties of a 4-dimensional spacetime from relationships between points of matter and energy. There is no absolute Newtonian space and time onto which the gravitational field (spacetime) is mapped. This relativistic description is what makes unifying gravity with the background-dependent Standard Model quantum field approach so difficult. To investigate this foundational problem between background dependency and independency, some investigators have taken a step back and looked beyond the descriptions and mathematical tools provided by the two incompatible approaches to glean the outlines of a deeper theory that is beyond quantum mechanics and general relativity themselves.

The basic idea is to describe all of the properties of quantum fields in terms of new objects that do not require a background coordinate framework, for example, to describe the properties of a field in terms of things other than space and time coordinates. General relativity is a covariant theory, which means that all physical quantities are described in terms of tensors. These tensors provide a way to state laws in a coordinate-independent way so that the form of the law remains unchanged (invariant). But what you have to do, to make this description background-independent is to describe all of the fields in the Standard Model in terms of quantities that reference only properties of the gravitational field, not some pre-existing spacetime that provides the coordinates. Generally, when you describe a quantum state, or the strength of the electromagnetic field, you do so in terms of their amplitudes at a particular spacetime point such as $A(x,y,z,t)$ for electromagnetism, while the general description of the field is given as a tensor A^μ, called a 4-vector. You now have to describe these fields in terms of the 4-dimensional spacetime provided by the gravitational

Carlo Rovelli is one of the architects of loop quantum gravity, which is attempting to create a fully relativistic theory of spacetime at the planck scale.

field itself. In general relativity, this field is described by the tensor $g_{\mu\nu}$. Symbolically, you might write something that looks like $A(g_{\mu\nu})$, but although this makes no mention of any underlying spacetime coordinate system, it is very hard to intuitively understand. One of the first of these background-independent theories for gravity was developed by Lee Smolin (1955–) and Carlo Rovelli (1956–) in 1994 and called Loop Quantum Gravity (LQG).

LEE SMOLIN

Born in 1955, Lee Smolin received his PhD in physics from Harvard University in 1979. He held many faculty positions at Yale, Pennsylvania State and Syracuse University before becoming a Visiting Scholar at the Princeton Institute for Advanced Study in 1995. Smolin had been interested in quantum gravity issues since the late 1980s and worked with Carlo Rovelli, Abhay Ashtekar (1949–) and Ted Jacobson (1954–) to formulate such a theory called Loop Quantum Gravity based on purely relativistic principles with no pre-existing spacetime. Smolin has also been an active popularizer, and investigator of both experimental and philosophical consequences of quantum gravity theory.

LQG was mathematically designed not to be dependent on a pre-existing spacetime as found in general relativity. Instead, this theory creates spacetime when the separations between points is large compared to the Planck scale of 10^{-33} cm. At the Planck scale, space is simply a collection of points that each have a Planck unit of volume. But instead of the points being connected by space-like intervals they are connected by relationships that can be thought of as transporting a quantum of Planck area to each point vertex. These relationships define what is called a spin network because the areas can be shown to form a spectrum similar to the angular momentum quantum numbers in atomic physics. Specifically, the area spectrum formula was found to have the specific form

$$A = 8\pi L^2 \sqrt{J(J+1)}$$

where L is the Planck length 10^{-33} cm and J is the spin quantum number (1,2,3, etc) of the segment connecting each Planck volume vertex point. A small piece of a spin network looks like the figure shown below.

The spin network in the figure shows that the total area 'transported' to a node is related to the sum of the quantum areas carried by each line segment that intersects the vertex. For example, in the accompanying figure, the vertex at the centre of the hexagon-like network is linked to the clockwise-labelled segments representing J = 1, 2, 4, 2, 1, 4. Using the area formula for each of the 6 segments, we see that the total area represented by the vertex volume is then $A = 8\pi L^2 (\sqrt{2} + \sqrt{6} + \sqrt{20} + \sqrt{6} + \sqrt{2} + \sqrt{20})$. The segment spins, J, cannot be arbitrary but have to follow specific rules based on the spin arithmetic developed by Roger Penrose (1931–) in his related theory of Twistors developed in 1971.

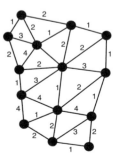

A simple spin network.

According to LQG, each spin network represents a specific quantum state of the gravitational field. By encircling a collection of nodes in a spin network, the sum of the nodes generates a specific volume of 3-dimensional space, and the segments that pierce this encircled path represent the surface area of the volume. It is the expectation of the LQG programme that at very large scales, the discrete formulae that define large spin networks will look increasingly more like the equations we use to define distances in the ordinary 3-dimensional space, and which are used as the background for the Standard Model.

Spin networks show promise in creating ordinary space starting from a completely non-spatial area. But what about time? The gravitational field is represented by a 4-dimensional spacetime, and to get this time-like extension, a succession of spin networks are related to each other by specific moves or alterations in which vertices are connected. The resulting 4-dimensional structure of sequentially altered spin networks is called a spin foam.

Thus far, although LQG provides a framework for describing gravitational fields from a purely relativistic perspective, there is no way as yet to move from a spin foam description to the larger-scale description of spacetime used in either standard quantum mechanics or classical general relativity. Also, there are as yet no features of spin networks and spin foams that can be used to represent any of the quantum fields in the Standard Model. However, in the area of black hole physics and information theory, LQG and superstring theory calculations do, when confronted with the same physical problem, lead to the same answers.

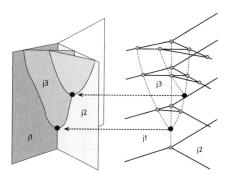

An example of a spin network changing in time in a series of three moves. The time-like axis is in the vertical direction.

Key Points

- Quantum mechanics, the Standard Model and superstring theory are all based on a pre-existing 4-dimensional spacetime and are called background-dependent theories.

- A true quantum theory of gravity must share the basic principle in general relativity that the theory is background independent so that 4-dimensional spacetime is created as a prediction of the theory and does not pre-exist.

- Currently only Loop Quantum Gravity is a background-independent quantum theory of spacetime.

- In LQG, spacetime dissolves into discrete Planck-scale volumes of space connected by relational rules that assign specific quantized areas to each quantum volume following specific quantum rules related to spin. These are called spin networks.

- A collection of spin networks organized by their changes leads to spin foams in 4 dimensions that resemble the properties of quantized spacetime.

CHAPTER 16
Antimatter

Since Dirac came up with the idea of antimatter, and Anderson discovered the electron's antiparticle the positron, the existence of antimatter has been something of a surprise to physicists. Quantum mechanics and the Standard Model exist in a symmetric state in which both forms of matter may be treated in exact equivalency. In the former case, all calculations of wave functions and quantum states apply the same to matter as to antimatter; in the later case, all of the 25 Standard Model particles have their antimatter analogues. But observationally, antimatter is difficult to produce, and there is no exact symmetry between matter and antimatter in the macroworld beyond the scale of atomic physics. In the universe at large, this leads to the thorny cosmological difficulty at the dawn of the origin of the universe when matter and antimatter were in direct physical contact, though now we clearly live in a universe that at all scales is dominated by matter.

The rules for antimatter are simple. When a particle and its antiparticle are paired together, the sum of all their quantum numbers must be identical to the vacuum state in which all quantum numbers sum to zero. An electron can have a charge of -e. Its antiparticle the positron can have a charge of +e. The sum of the paired charges is zero. For the proton, it is composed of three quarks: two up quarks with q = +2/3e and one down quark with q = -1/3e. The antiproton consists of two anti-up quarks with q = -2/3e and one antidown quark with q = +1/3e. Similarly, the antineutron consists of two antidown quarks with q = +1/3e and one antiup quark with q = -2/3e. Photons carry no charge and are their own antiparticles, while gluons carry colour charge such that the antiparticle to the red-antiblue gluon is the blue-antired gluon.

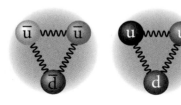

Quark structure of the proton (right) and the antiproton (left).

When a particle and its antiparticle combine, the net quantum numbers must be zero as for the vacuum state. This means that although the sum of the charges naturally equal zero, so too must the net energy and momentum because the vacuum state has zero rest mass. To make this happen, the rest mass energy $E = mc^2$ must be carried off by photons, but because photons carry momentum, this requires that two photons be generated, each travelling in exactly opposite directions so that their total momentum sums to zero.

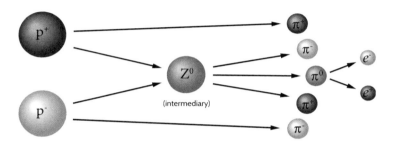

When a proton and anti-proton annihilate, they produce a spray of new particles in the ball of pure energy that results.

Electron-positron annihilation is rather straightforward. They combine to produce a quantum system with no net quantum numbers, and two photons are generated that travel in opposite directions to conserve momentum, and each carries an energy equal to $E = mc^2$ or $511,000$ eV/c^2 in the usual units of atomic physics. For baryons and antibaryons that consist of quarks, the process is much more complex. For protons and antiprotons, their quark and anti-quark constituents temporarily form a quark-antiquark plasma with attendant gluons and antigluons. This merged system then forms temporary pairs of mesons and antimesons which themselves consist of quark-antiquark pairs. These mesons then decay into electrons, positrons and gamma rays. Theoretically, as many as 13 intermediate meson states can occur, with the largest number detected being 9. The generated mesons then leave the annihilation region and subsequently decay into their respective electrons, positrons and gamma rays at the decay rates appropriate to each meson. Antiprotons and ordinary neutrons can also annihilate, because each consists of three quarks whose quantum numbers can sum to zero. Typically, a proton-antiproton annihilation results in the production of two p° mesons or two K° mesons. However, annihilation into charge conjugated pairs such as $\pi^+\pi^-$ and K^+K^- also occur about one per cent of the time. The energy of the annihilation is shared by the end products according to their masses such that photons and electrons with the least rest masses carry off the lion's share of the energy. Other mesons are also possible and in fact the investigation and detection of new types of mesons is made easier during proton-antiproton collisions, hence the desirability of constructing accelerator laboratories in which beams of protons and antiprotons are collided, such as the Tevatron at Fermilab in Illinois and the Large Hadron Collider at CERN.

The fact that we live in a matter-dominated universe, and not one symmetric in matter and antimatter, suggests that there exists some fundamental mechanism which breaks the symmetry between these two otherwise equal states. Cosmological measurements suggest that this asymmetry needs to be only about one part in ten billion favouring matter, in order for the universe to appear as it does and be dominated by matter.

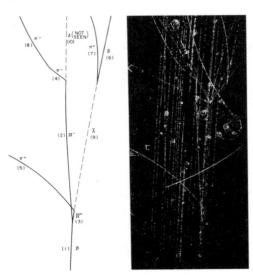

Bubble chamber photograph of a proton-antiproton collision that generates several π mesons together with other exotic mesons.

One way in which this can occur is if there is a mechanism that slightly favours matter states over

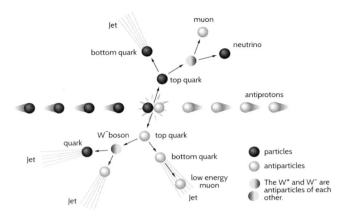

Top Quark – Antitop Quark Event.

antimatter states as the end products of particle decay. For example, a neutron decays after 881 seconds into an electron (q = -e), a proton (q = +e) and an antineutrino. An antineutron decays after 881 seconds into an antiproton (q = -e), an antielectron (q = +e) and a neutrino. Let's consider what would follow if the weak interaction mediated by the W^+, W^- and Z^0 particles favoured the neutron decay channel over the antineutron decay channel: even though equal numbers of neutrons and antineutrons existed, their decays would lead to a slight overpopulation of protons and electrons and hence produce a net baryon/antibaryon asymmetry. Is there any evidence for such a decay asymmetry? In fact, there is, but to explore this we must investigate a new symmetry of Nature called CPT Invariance.

THE CPT SYMMETRY

Charge (C), Parity (P) and Time (T) reversal are combined so that if we placed any particle facing a mirror and investigated its mirror-world behaviour, the following should occur. Change the charge of the particle from + q to - q, change the direction of spin – called its parity or handedness from clockwise (spin up) to anticlockwise (spin down) – and the direction of time from +T to -T (equivalent to reversing the momentum of a particle from +P to -P) and the resulting physics of its mirror-world twin will look identical. In other words, the interaction is symmetric and obeys CPT Invariance. This idea was developed in 1951 by Julian Schwinger (1918–1994), one of the developers of QED. No known physical law violates CPT symmetry, and in fact this is a symmetry that all theories in physics must pass in order to be seriously considered for further analysis.

We can examine CPT symmetry by breaking it up into parts. The first of these studied in detail was Parity, which was believed to be conserved by the mid 1950s. For example, if you looked at a clock in a mirror, the hands of the two clocks would move towards each other and consequently the spin of the normal particle going clockwise would be reflected in its mirrored image by the hands (spin) rotating anticlockwise. If you changed the parity of the mirrored clock so that the hands moved clockwise,

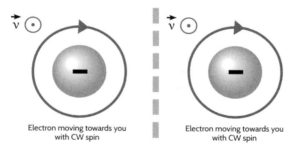

Electron moving towards you with CW spin   Electron moving towards you with CW spin

Example of CPT symmetry.

the mirrored image would still not look like the normal clock. That means that parity is not a symmetry of the system. In other words, changing the parity of a system changes how it behaves. Experiments conducted in the 1950s showed that the strong and electromagnetic forces did preserve parity symmetry, but the weak interaction did not. Experiments in 1957 by Chien-Shiung Wu (1912–1997) at the National Bureau of Standards using cobalt-60 nuclei did show that parity was violated in beta decays. There should be no polarization difference between gamma rays emitted by nuclei spinning in a fixed direction, but in fact 40% were emitted in one direction while 60% were emitted in the other. Parity conservation required that there be no difference.

If parity conservation is violated, then perhaps it can be balanced by changing the charge so that the combined CP remains a symmetry. In other words, flip the coordinate system 'upside down' by changing its parity, but simultaneously change a particle to its antiparticle, and then the mirror-world process should look the same. In 1964, an investigation of K meson decay by James Cronin and Val Fitch showed that CP is violated. They found that neutral K mesons can transform into their antiparticles and vice versa – but not with exactly equal probability. The neutral K° comes in two forms: K°_{Long} has a lifetime of 5×10^{-8} seconds and the K°_{Short} has a lifetime of 9×10^{-11} seconds. The K_L mostly decays into three π mesons while the K_S mostly decays into two π mesons. Because both Ks have an intrinsic CP of -1, and the pions have CP = -1, we see that the decay of the K_L preserves CP because for the three π mesons $(-1)(-1)(-1) = -1$, but the K_S violates it because for its two π mesons $(-1)(-1) = +1$. This is also

JAMES CRONIN

James Cronin (1931–2016) obtained his Doctorate in 1955 at the University of Chicago, where his thesis advisors included such Nobel luminaries as Enrico Fermi, Murray Gell-Mann and Subrahmanyan Chandrasekhar. He moved to the Brookhaven National Laboratory where the new Cosmotron accelerator was being used and began a study of parity violation in nuclear reactions. In 1958 he moved to Bevatron at the University of California, Berkeley, where he and Val Fitch (1923–2015) studied the decay of the neutral K meson (or kaon), in which they discovered CP violation in 1964. This research led to their Nobel Prize award in 1980. During the last years of his professional life, he investigated high-energy cosmic rays and their sources in space.

James Cronin (top centre) was instrumental in investigation CP violation in the decay of K-mesons.

true for the decays of the K_L and K_S antiparticles, so that CP invariance is not preserved in K meson decay under the weak interaction. The decay of the K meson was the only example of CP violation until 2001, when it was found in the decays of the B mesons, and then in 2019 in D mesons.

In the Standard Model, K, B and D mesons allow decays involving matter particles to proceed at a different rate than for their antiparticles, so this mechanism seems a hopeful answer as to why the universe is matter-dominated. Unfortunately, the difference in rate is much too small to account for the observed cosmological imbalance, which is estimated to be 1 in 10 billion. In other words, the cosmological imbalance has to be such that 10 billion + 1 matter particles were produced compared to 10 billion antimatter particles, while the weak CP violation allows for only about one millionth of the required CP violation to solve the cosmic antimatter problem. This remains a daunting and unsolved problem in physics that seems to be beyond the Standard Model to understand.

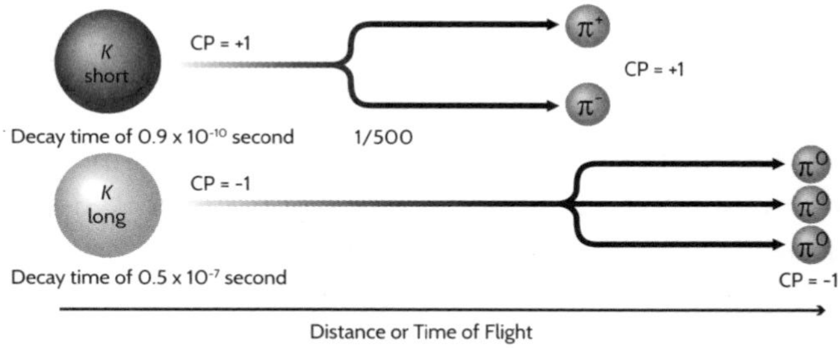

CP violation in the decay of the two K-mesons leads to two different outcomes.

Key Points

- Antimatter resembles ordinary matter except that the charges of its constituent particles are reversed.

- Antimatter electrons, called positrons, have a positive electric charge. Antimatter quarks have opposite electric charges and allow particles such as protons and neutrons to have antimatter partners.

- Although a neutron has an electric charge of zero, its antimatter version is constructed from three antimatter quarks whose opposite charges sum to a zero net charge.

- CP violation in the weak interaction proves that Nature slightly favours matter over antimatter states in some nuclear interactions.

- The one-to-one billion overabundance of matter over antimatter in our universe cannot be understood in terms of the known weak interaction CP violation because the cosmic overabundance is about one million times too large.

Further Reading

CHAPTER 1
The Dawn of Atomic Physics
Segrè, Emilio. *From Falling Bodies to Radio Waves: Classical Physicists and Their Discoveries.* Dover Publications (1984).

CHAPTER 2
Quantization
Segrè, Emilio. *From X-rays to Quarks: Modern Physicists and Their Discoveries.* Dover Publications (1984).

CHAPTER 5
Quantum Measurement
Becker, Adam. *What is Real? The Unfinished Quest for the Meaning of Quantum Physics.* John Murray (2018).

CHAPTER 9
Quantum Field Theory
Feynman, Richard. *QED: The Strange Theory of Light and Matter.* Princeton University Press (1985).

CHAPTER 10
The Standard Model
Still, Ben. *Particle Physics Brick by Brick.* Cassell (2017).

CHAPTER 11
Supersymmetry
Greene, Brian. *The Elegant Universe: Superstrings, Hidden Dimensions and the Quest for the Ultimate Theory.* Vintage (2000).

CHAPTER 12
String Theory
Randall, Lisa. *Warped Passages: Unraveling the Mysteries of the Universe's Hidden Dimensions.* Harper Collins (2005).

CHAPTER 13
Quantum Gravity
Smolin, Lee. *Three Roads to Quantum Gravity.* Basic Books (2001).

CHAPTER 15
Loop Quantum Gravity
Rovelli, Carlo. *Reality Is Not What It Seems: The Journey to Quantum Gravity.* Penguin Random House (2018).

CHAPTER 16
Antimatter
Close, Frank. *Antimatter.* Oxford Landmark Science (2010).

Glossary

Anthropic Principle The idea that the natural constants of the universe are selected by virtue of their allowing sentient life to measure them to have the values they have.

Antimatter A particle identical in mass and spin to ordinary matter except that its charge value is of opposite sign.

Boson A particle with a spin value of 0, 1 or 2, such as photons and gluons.

Brane A subspace of 11-dimensional spacetime consisting of three or more dimensions.

Calabi–Yau manifold The space-like manifold consisting of the additional dimensions to spacetime needed to form the basis for GUT theory. Its shape and geometry presumably determines the properties of the Standard Model particles.

Dirac hole A state of the quantum vacuum that is of negative rest mass energy compared to normal electrons.

Fermion A particle with a spin value of ½ such as an electron, quark or neutrino.

Feynman diagram A diagram used to identify the factors that appear in quantum field calculations.

GeV An energy unit equal to 1 billion electron volts or 1.6×10^{-10} Joules.

Gluon The massless particle that mediates the strong nuclear force between quarks.

GUT Grand Unification Theory – a class of theories that unify the strong, weak and electromagnetic forces.

Hidden dimensions The proposal that 4-dimensional spacetime also contains quantum-scale additional space-like dimensions, perhaps as many as 7.

Higgs boson The fundamental particle responsible for providing masses to elementary particles and breaking the symmetries between the forces that act upon them, specifically the weak and electro-magnetic forces.

Landscape A presumably mathematical space analogous to Hilbert space in quantum mechanics, consisting of all possible spacetimes allowed by string theory via the compactification of its Calabi–Yau manifolds.

Lattice gauge theory The technique used in strong interaction calculations in which spacetime is a lattice connecting quarks at the vertices.

Lepton The family of light fermions including the electron, muon, tauon and associated neutrinos.

Loop Quantum Gravity A proposed theory that quantizes gravity by using relational principles at the

Planck scale to create spacetime from more elementary objects.

M-theory A theory of 11 dimensions which unifies the five possible superstring theories.

Multiverse The proposed arena consisting of innumerable disconnected spacetimes, each with its own compliment of equivalent Standard Models in which our universe is only one of many possibilities. Related to the idea of the Landscape.

Neutralino A particle predicted by extending the Standard Model using supersymmetry, and which has the properties to explain cosmological dark matter.

Observer Participation The idea related to Local Realism that the act of observing a quantum system brings it into existence as a 'real' particle with specific measurable properties.

Photon The elementary quantum particle of the electromagnetic field.

Physical vacuum The properties of classical empty space based upon quantum principles, especially Heisenberg's Uncertainty Principle.

Planck Scale A scale of time, distance, mass and energy defined in terms of the three fundamental constants: the speed of light, the constant of gravity and Planck's Constant at which quantum changes in gravity and spacetime should occur.

Positron The antiparticle to the electron with a positive electric charge and identical mass.

Quantum A discrete feature of physical systems resembling individual grains of sand on a beach.

Quantum chromodynamics The relativistic quantum theory describing the strong interaction in terms of gluons and quarks.

Quantum electrodynamics The relativistic quantum theory of the electromagnetic interaction in terms of electrons and photons.

Quantum field theory The theory of physical particles and fields based upon relativity and the quantization of matter and energy.

Quantum gravity The family of theories in which the gravitational force and spacetime are described in terms of quantum phenomena.

Quark The family of six elementary particles constituting the elementary particles that experience the strong force.

Renormalization A mathematical technique used in relativistic quantum field theory to eliminate infinite answers in calculations of finite observable processes.

Spacetime The term used to describe the combination of the one dimension of time and the three dimensions of space in special and general relativity. It is the physical basis for the gravitational field.

Spin foam The quantized structure of 4-dimensional spacetime according to Loop Quantum Gravity.

Spin network The quantized structure of 3-dimensional space according to Loop Quantum Gravity.

Spontaneous emission The ability of a quantum system involving photons to suddenly change from one state to another despite these states each being stable.

Spontaneous symmetry breaking The ability for a system of fields and particles to transition from one set of symmetry principles to another set of reduced symmetry.

Standard Model The current experimental and mathematical model for the strong, weak and electromagnetic forces, consisting of 25 fundamental particles grouped into two families, with fermions representing matter particles and bosons representing the force-mediating particles.

Superfield A proposed unification of fermion and boson fields into a new mathematical field that has the properties of both.

Supergravity A proposed description of gravity as a spin-2 quantum field.

Superspace An arena similar to Hilbert Space in ordinary quantum mechanics in which each point is a complete quantum state of our entire universe, and the evolution from state to state is determined by a Schrödinger-like equation called the Wheeler–DeWitt equation.

Superstring The proposed description of the structure of elementary particles represented by a 1-dimensional string vibrating in 10 dimensions, and for which

supersymmetry allows these strings to share the properties in the Standard Model.

Symmetry group The set of operations, such as elementary rotations, forming a mathematical group in which combinations of these operations leave some property of an object invariant. For example, rotating a cube clockwise by 90 degrees still leaves the final cube looking the same as the original cube.

The Bulk The term used to describe the 11-dimensional spacetime of M-theory.

Theory of Everything The term used to describe a theory that accounts for all of the properties of the Standard Model as well as the quantum properties of gravity and spacetime.

Virtual particle A particle in a negative-energy state, which is required in current quantum field theories of the Standard Model interactions.

Yukawa particle A particle such as a gluon that carries its own charge as well as a non-zero mass, which determines its range in space according to an exponential formula.

Zero-point fluctuation The quantum variation in the physical vacuum due to the appearance and disappearance of virtual particles and fields.

Index

Picture Credits

t = top, b = bottom, l = left, r = right

Caltech: 210 (MIT/LIGO)

CERN: 79

Getty Images: 18, 58, 62, 104, 107t, 136, 174, 180

NASA: 205, 211

Public Domain: 13, 14, 17, 19, 20, 21t, 21b, 22, 26, 32, 39, 43, 44, 45, 49, 51, 52, 53t, 53b, 59, 60, 63, 71, 72, 73, 74, 75, 78, 85t, 85b, 86, 94, 98t, 98b, 100, 107b, 111, 112, 116, 119, 138, 139, 140, 144, 147, 148, 159, 160, 162, 176, 178, 185, 188, 191, 193, 204, 208, 209, 212, 218, 220, 221, 224, 225, 226, 227, 228, 229, 230

Science Photo Library: 15, 96, 126, 189, 207, 213

Shutterstock: 42, 157, 169, 179

Wellcome Collection: 11, 16, 30, 37